THE WORLD OF COPERNICUS

THE WORLD OF COPERNICUS

(Originally published as 'Sun, Stand Thou Still')

by
Angus Armitage
M.Sc.(Lond.), F.R.A.S.

Republished by EP Publishing Limited
1972

This edition published 1972 by
EP Publishing Limited, East Ardsley,
Wakefield, England, by kind
permission of the copyright holders, Abelard-Schuman, Ltd., London

Reprinted 1976

This edition offset by kind permission
of New American Library, New York

Originally published under the title
'Sun, Stand Thou Still' by Henry Schuman Inc., 1947
and as 'The World of Copernicus' by Mentor Books, 1951

ISBN 0 85409 737 6

Please address all enquiries to EP Publishing Limited
(address as above)

Reprinted by The Scolar Press Ltd.,
Ilkley, West Yorkshire, U.K.

CONTENTS

PART THREE

THE TRIUMPH OF THE COPERNICAN THEORY

Illustrations

Plates

Preface

In the following pages I have tried to tell the personal story, and to convey some of the historical motives, at the back of the great sixteenth-century revolution in man's outlook on the Universe. I hope the book may be of interest, not only to astronomers and to students of the history of science, but also to the general reader desirous of tracing back to their sources some of the scientific ideas which have helped to shape modern thought.

It is impossible to treat a great thinker like Copernicus in isolation from the intellectual tradition which he inherited, or from the age which felt the first impact of his ideas. I have therefore felt obliged to devote a considerable proportion of my allotted space to explaining the origin of the situation in which Copernicus played his decisive part. The first dozen chapters, making up Part One, have accordingly been planned to show how the old-time system of the world, and the principles on which Copernicus refashioned it, were alike conditioned by ideas which had developed from ancient times. On the other hand, the last dozen chapters, constituting Part Three, are intended to round off the present study by sketching the transformation and triumph of the Copernican theory during the century following its publication. I hope therefore that the book may be found to provide not only an adequate account of Copernicus and his achievements, but also an introduction to the history of astronomy in its wider aspects.

I desire to thank the British Museum for the use I have been allowed to make of the Reading Room during the preparation of this book; also for kind permission to reproduce the illustrations forming Figs. 8, 10, 16 and Plates II, III, IV, V, VI from works in the possession of the Museum. The Mundus Publishing Association Ltd. kindly provided me with the photographs for Plates I and VII; and I am grateful to Professor Charles Singer and the Clarendon Press, Oxford, for leave to reproduce Fig. 8.

While I assume full responsibility for any errors or omissions in this somewhat experimental book, I feel deeply indebted to Dr. Edward Rosen of the College of the City of New York who kindly read through the typescript and made valuable criticisms and suggestions. Lastly, I would offer my thanks to Miss Doris Meyer for drawing the diagrams and the map, and to the Publisher, Mr. Henry Schuman, for his advice and encouragement throughout the preparation of the book.

December, 1946

curiosity all down the ages. The solution of this problem was, in fact, his lifework.

In the days when Copernicus lived, everybody believed that the *Earth* was fixed at the center of the Universe, and that the heavenly bodies, including the Sun, all went round the Earth in circles. Copernicus, on the other hand, maintained that the *Sun* was fixed at the center of all things, and that the Earth, itself a heavenly body, revolved round the Sun once in a year. The other neighboring bodies, he supposed, also made circuits round the Sun, not, indeed, in a year, but each in a period of its own. He further asserted that the Earth turned completely round on itself once in a day.

We are all convinced today that the solution put forward by Copernicus was on the right lines. But his views were not accepted immediately he put them forward. For other thinkers, long before his time, had tackled the same problem. In fact, we must emphasize right from the start that the great contribution made by Copernicus to the rise of modern thought was of the nature of a *choice* rather than of a *discovery* of something new. One or two of the earlier thinkers had, indeed, proposed the same solution as he proposed. But the great majority of them took the common-sense view that the Earth is fixed at the center of the Universe just because that is how things appear to us. And so, as we have said, that was the belief that prevailed in the days of Copernicus. Around that belief there had been built up a great system of scientific and religious thought which was generally accepted throughout Christendom.

We cannot hope to understand the life and achievements of Copernicus unless we take account of that system of thought. For on the one hand it was the tradition in which he was brought up and from which he started out, and whose prejudices he never entirely outgrew. He was indebted to it for his knowledge of the facts which he was going to interpret in a new way, for his methods of work, and for the ideals which he set himself to achieve. On the other hand, Copernicus was also a rebel against the traditional view of the world, whose supporters long continued to fight against his ideas. In fact it was not until about a century and a half after the death of Copernicus in 1543 that the Earth-centered Universe of the ancients yielded place to the rival Sun-centered Universe which he had conceived.

Hence this book falls naturally into three parts. We have first to explain how the special problems that concern us here arose in the wider setting of the human epic, and how they were treated down to the time when Copernicus tackled them. This will occupy Chapters 2 to 12. Next, in Chapters 13 to 33 we shall tell the story of the life of Copernicus, so far as the particulars have come down to us. We shall also give a general account of the basic ideas set forth in his historic book. Lastly, in Chapters 34 to 45, we shall follow the principal stages in the establishment of the Copernican system of the world down to the age of Newton, noting, in conclusion, its wider significance for our own day.

2. The Human Adventure

Man has been living on the Earth for upwards of a quarter of a million years. During almost the whole of that time, his life was one continual struggle to keep himself alive and to rear his young. Human development has not everywhere followed the same uniform course. But, generally speaking, man began by feeding on the fruits and roots that grew around him, and that could be had for the gathering. Or he hunted wild beasts for food to eat and for skins to cover his nakedness. He lived in caves and traveled in small bands of his kind with no more belongings than he could easily carry.

Then some time about ten thousand years ago, in certain parts of the world, man began to find out how to secure the food he needed without having to use up all his time and strength in doing it. He learned how to grow plants good to eat, and how to keep tame animals to provide him with fresh meat and milk. Thus, over large areas of the Earth's surface, he exchanged the wandering existence of the hunter for the more settled life of the shepherd or farmer.

This marked the beginning of man's rise from savagery to the position he holds in the world today. For shepherds and farmers can produce more food than they themselves need. There is thus a *surplus* of food; and this can go to feed other people who, because they are not obliged to feed themselves, can spend their time and strength doing or making other things. It is these "other things" which have done so much to enrich man's life, and to raise it above a merely animal level.

Foremost among these things, we must reckon tools, made at first of stone, later of metals extracted from natural ores by fire. Tools enabled man to do many things that he could not have done with his bare hands. At the same time, he learned to employ mechanical aids such as the wheel and the lever. He began to use materials, making bricks and building houses to live in, making pottery and glass vessels to hold his food and drink. He learned to spin thread and to weave cloth, to work in leather, bone and metal, and to practice many other arts. Some men devoted themselves to particular trades, such as that of the blacksmith, so as to acquire greater skill, in this way. One community would obtain anything it lacked from another community by peaceful exchange or by force. It became profitable to set men free from food-producing to engage in trade and war.

Among the most effective tools of early man, we must include writing instruments. Writing has enabled him to accumulate knowledge beyond what he could merely remember and pass on to the coming generation by word of mouth. And writing is an aid to reckoning; without it, mathematics, and the sciences based upon it, could scarcely have come into being at all.

However, at some early stage of his development, man seems to have become conscious of the mystery of the world around him. He began to have a sense of haunting presences, calling him to account, but perhaps willing to help him in his struggle for life. Man pictured the various forces of nature—storm and sunshine and the life of the growing grain—as beings like himself, only much more powerful. He felt that his welfare depended upon their goodwill. And he believed that if he performed certain magic rites, these beings could be persuaded or compelled to bless his crops and flocks and other personal concerns.

This belief led to another class of men being set free from food-producing for special duties. Besides the craftsmen and the merchants, there were the priests whose task it was to manage the relations of the community with the unseen world. It was the priests, too, who first had the leisure to make up elaborate stories with which to satisfy men's curiosity about the world around them. "Man doth not live by bread only"; and out of the crude magic of an earlier day there grew up increasingly refined forms of natural religion. These satisfied, in some measure, the spiritual needs of man's nature, until, in the fullness of time, they gave place to something better.

However, so far as the schemes for controlling events by magic were concerned, men have learned from bitter experience that nature does not work in that way. Things do not happen in the world as if at the bidding of capricious sprites, but in an orderly manner, as if in deference to fixed laws which have to be obeyed. Night follows day; the seasons pass in due succession from seedtime to harvest; the events of human and animal life form a regular routine. The discovery of this essential orderliness of nature marks the beginning of what we call science. Once men have reached the scientific stage of development, they realize that success in living does not depend upon coaxing or forcing nature to do what we want. It depends upon understanding nature's laws, and in making use of them to serve human purposes. That is the principle underlying all the great inventions —steamships, airplanes, radio and so forth—which loom so large in the world today.

Besides ministering to our comfort, Science also serves to satisfy certain needs of the human spirit. It helps us to understand the world and to feel at home in it. We are distressed by disorder, and always try to arrange in order the things with which we have to deal, whether they are the affairs of a nation, the books in a library, or our own ideas. Science satisfies us because it shows us that, behind the transient and confused pageant of nature, there is a permanent and orderly reality. And it was in the sky that this order was first revealed to men on a vast and spectacular scale.

3. The Face of the Sky

When men first began to take notice of the world around them they must have been tremendously impressed by what they saw in the sky. It is no wonder that they made gods of the Sun and Moon, and worshipped the host of heaven. But soon they began asking what were the rules according to which the heavenly bodies move, and why they behave as they do. All down the ages men have tried to find answers to these questions. In the following pages we have to tell the story of one man who gave the most satisfying answer of any down to his time. His work served as a foundation upon which a great part of the modern science of the heavenly bodies has been built.

The lapse of ages makes but little difference to the face of the sky. It appears the same to us today as it did to our forefathers. What they knew about the heavens, we can easily rediscover for ourselves. And it will help us to understand the matters with which we shall be concerned in the following pages if, at the outset, we get a clear idea of what astronomy is all about.

Let us imagine, then, that we are standing upon some hill or watchtower viewing the heavens on a cloudless night. We see the stars as points of light against the dark blue background of the sky. We notice that they differ one from another in brightness, and that they are irregularly arranged so as to form star-groups which are called *constellations,* and which reappear in the sky night after night. Perhaps we learned to recognize some of these in childhood—the Dipper, Orion or the Pleiades. They received their names in early times after the heroes, monsters or familiar objects they were thought to resemble.

If we continued to watch the sky for some time, we should discover that the stars generally rise in the east, pass across the sky, and set in the west, exactly as we see the Sun do every day. If we observed very carefully, we should discover that the stars behave as if they were fixed to the inner surface of a huge sphere which is slowly turning round with us at the center (Fig. I). The earth on

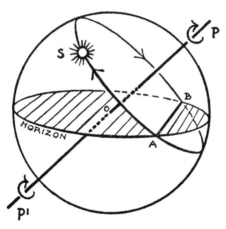

FIG. 1. THE SPHERE OF THE SKY

The apparent turning of the sphere of the sky about its two opposite *poles* P and P¹ makes the star S appear to the observer O to rise above the *horizon* at A and to set at B.

which we are standing seems roughly like a flat surface stretching out in all directions and cutting the sphere of the sky in a great circle which we call the *horizon*. Half of the sphere is under our feet and hidden from view by the Earth; and a star comes into view when the turning of the sphere brings it above the horizon. The sphere of the sky makes rather more than one complete turn in one day, so that each star rises about four minutes earlier than it did the day before. That explains why we do not always see the same stars at the same time of night throughout the year.

The stars provide us with a permanent background against which we can discover and measure the motions of other bodies. For example, if we observe the Moon night after night, we notice that she is near different stars on different nights. In fact, while she shares with the stars in their daily rising and setting, she also describes a complete circuit of the heavens against the background of stars, taking about a month to do it.

The stars fill the sky in the daytime as well as at night; but we cannot see them then because of the glare of the Sun. If, however, we could see the Sun against the background of stars, we should discover that he too, like the Moon, travels slowly round the heavens, completing his circuit in one year. This causes him to lag a little behind the stars in the daily motion of rising and setting. And as we measure our day by the Sun and not by the stars, this lagging of the Sun explains why the stars, as we saw, appear to rise four minutes earlier each day.

If we kept up our watch on the sky night after night, we should notice several objects that look like bright stars, but are not. They differ from stars by traveling slowly and rather irregularly round the sky from one constellation to another. These bodies, like the Sun and Moon, ordinarily move slowly from west to east, completing the circuit of the sky each in its own period of revolution. But sometimes they stop and travel a little way from east to west before resuming their eastward journey. These are the "wandering stars" or *planets*. The ancients recognized five of them; and they named them, as we still do, after their divinities: Mercury, Venus, Mars, Jupiter and Saturn.

This was about as much as our primitive forefathers were able to find out concerning the heavenly bodies. And it roughly represents the extent of astronomical knowledge among savages in the world today.

Let us take one more look at the discoveries we have just described. We notice that they are almost all concerned with events that happen over and over again at regular intervals of time. The daily rising of the Sun is an obvious example; we have had other instances of the same sort of thing in the monthly circuit of the Moon and in the yearly circuit of the Sun. We call such an event, repeated at regular intervals, a *recurrence;* and the interval after which it happens again we call its *period.*

Early astronomers were very interested in such recurrences, because they served, and still serve, to measure the lapse of time. In the beginning there was no such thing as a calendar; and without a calendar it would be impossible to know exactly when to sow the crops upon which the life of the community depended. It would be impossible to keep historical records, or to celebrate religious festivals at the right time every year. So early men measured time by the motion of the Moon, and her change of appearance as she passed from new moon to full and back to new again. That is, they reckoned the lapse of time in *months.* Or they made the same sort of use of the Sun, and reckoned time, as we still do, in *years.* The hunting tribes, who liked to choose moonlight nights for their expeditions, generally used a calendar based on the month. But the peoples who lived by agriculture based their calendar on the year, the period of recurrence of the seasons upon which the growth of the crops depends.

4. The First Astronomers

To discover the beginnings out of which modern astronomy has grown, we must look back to the ancient civilizations which sprang up beside the great rivers of the Middle East—the Nile and the Tigris-Euphrates. In rainless Egypt the skies are cloudless almost all the year round, and the stars shine forth with great brilliance. Such conditions naturally favored the growth of astronomy.

But it is especially among the Babylonians, who lived in the country we now call Iraq, that we shall find the earliest traces of a true science of the heavenly bodies.

Civilization flourished there as early as 3000 B.C. The inhabitants lived in walled cities the ruins of which can still be seen. They were in the habit of writing, or recording their observations, by making marks on tablets of soft clay, which were later dried or baked to form tiles. Many of these tiles have survived in the ruins of the old cities. About a century ago, students found out how to read them, and that is how we have come to know what Babylonian astronomy was like.

The observation of the heavens used to be carried on by the priests. They scanned the sky from watchtowers called *Ziggurats* which were built near the temples of the great cities. The priests were chiefly interested in observing the new moon, whose first appearance in the west marked the beginning of another month. They also followed the movements of the planets, noting the times when they passed near to each other or to bright stars. They looked out for comets, the "hairy stars" whose chance appearances have struck terror to the hearts of simple people all down the ages. They also took account of eclipses—the darkening of the Sun or Moon—which occur whenever the Moon passes between us and the Sun, or when the Earth's shadow is cast on the Moon.

The priests did not make these observations out of idle curiosity, nor, at first, through scientific interest. They made them in the belief that it was possible to foretell the future by taking note of what went on in the heavens. An eclipse of the Sun, or the appearance of a comet, might mean the near approach of war, famine or pestilence. It was the duty of the priests to warn the king or the citizens of what was in store. This belief that earthly events are foretold by signs in the heavens later developed into a kind of fortune-telling called astrology. It was believed that the heavenly bodies exerted a great influence upon the lives and the fortunes of every human being. It was even thought that a person's whole career depended upon the positions of the planets in the sky at the moment of his birth. We now know that this is an empty superstition, though some people still believe in it. The Babylonians had far more excuse for doing so. For they had no idea which events in the world were related to each other as cause and effect. They could only make guesses and trials in the hope of discovering the real connections between things. Moreover, even they largely outgrew astrology and became increasingly interested in the true science of astronomy. Let us look at some of their discoveries.

They made use of the Moon to measure their time, which they reckoned in months. They noticed that the Sun, Moon and planets, as they slowly revolve round the heavens, always keep within a narrow strip of sky extending like a belt right round the middle of the sphere of the stars. The Babylonians divided up this planetary belt into twelve equal parts; and they called each of the twelve divisions after a nearby constellation. Following them, the Greeks called the belt the *Zodiac,* and the twelve divisions, the *Signs of the Zodiac.*

The Sun travels through the twelve signs in one year. If he moved at a perfectly steady rate he would traverse each Sign in exactly the same time. But the Babylonians found that the Sun, Moon and planets did not move round the Zodiac at perfectly steady speeds. They drew up tables showing how the speeds varied from one Sign to another. That was the beginning of the *planetary theories* of which we shall learn more in later chapters.

The Babylonians tried to picture to themselves what the whole Universe was like. They thought of the Earth as a circular island rising towards the center into a huge mountain, and surrounded by a ring of sea. Beyond the sea was a circular range of mountains forming the boundary of the world and supporting the sky, which they thought of as a *Firmament,* or hollow half-sphere of solid material. There were doors in the sky to east and west through which the heavenly bodies passed at their rising and setting. There was water under the Earth which kept the springs flowing, and "waters above the Firmament" (about which we read in *Genesis*) which fell down as rain.

The only other people of those early times who can be compared to the Babylonians as astronomers were the Egyptians. But their achievements in astronomy were much less important. They grouped the stars into constellations and distinguished the planets, but their picture of the Universe was a very crude one. They thought of the sky as a cow with feet resting on the Earth, or as a woman supporting herself on hands and feet. Or they imagined it as a great sheet of water upon which the heavenly bodies sailed in boats. The Sun was for them a god sailing thus on the ocean of the sky, and descending at night beneath the Earth to visit the abode of the dead.

5. The Coming of the Greeks

About three thousand years ago there descended from the heart of Europe a group of primitive peoples who settled in the country we now call Greece. Some of them took to the sea and spread out eastward to settle on the islands of the Aegean, or on the western coast of Asia Minor. Others sailed westward along the Mediterranean to found colonies in southern Italy and Sicily. These people called themselves the Hellenes; we know them as the Greeks.

They settled down in small independent states, little more than cities, cut off from one another by barriers of mountain or sea. Despite the ties of language and religion which bound them together, these "city-states" engaged in almost incessant warfare with one another, until about 350 B.C., when Greece was made part of the military empire of Macedon. This empire, under Alexander the Great, proceeded to conquer a large part of the Middle East, and to spread Greek civilization as far as India.

The Greeks occupy an important place in our story. The whole idea of a science of the stars, and, indeed, all reasonable thought about life and the world practically started with them. It was the Greeks who first really tackled the astronomical problem which Copernicus was later to take up afresh and to carry a great stride nearer its solution.

The Greeks were well fitted by nature, and well placed, for the great part they had to play in moulding our civilization. In the first place they were a people whose minds were turned outwards to the world in eager curiosity. Their amazing skill in observation is shown in their paintings and carvings of the forms of men and animals. Again, the Greeks were extremely well informed about what other people had done. They flourished in a brilliant period of the world's history. They took over much of the culture of the highly civilized people who had lived before them in Greece and the Aegean. Perhaps it was from these earlier races that they acquired some of their artistic skill. On the other hand, the Greeks who had settled in Asia Minor were in close touch with the civilizations of Babylonia and Egypt, and, later, of Persia and, perhaps, India. The work on astronomy of which we spoke in the last chapter influenced the Greeks, and was latterly influenced by them.

Upon what they had thus observed for themselves or learned from others, the Greeks brought their bold, self-reliant minds to bear. They thought themselves quite capable of finding out how the Universe worked. As citizens of free states, the Greeks were accustomed to govern themselves by holding public meetings. They used to listen with open minds to the arguments of orators speaking for or against any proposal, and then they would decide the matter by a vote. They followed much the same method in trying to make up their minds what they should believe about nature, including human nature. The Greeks were, in fact, the first people to use argument deliberately as an instrument for discovering truth. They would start an argument from truths already established. They would carry it on according to sound methods of reasoning. And they would follow it through to its natural conclusion, however unexpected or even unwelcome that conclusion might be.

The religion of the Greeks never developed to the stage where it had a serious alternative view of the world to put forward which could conflict with the view to which they were led by common sense. They had no Bible or creed or powerful priesthood opposed to new ideas. Hence there was almost no persecution of the kind we shall meet with in the later pages of this book.

A good illustration of this Greek method of discovery by argument is afforded by the geometry of Euclid, who lived about 300 B.C. Euclid's book has been the basis of all the teaching of elementary geometry for some two thousand years, so nearly everyone has some familiarity with it. Euclid starts from simple and mostly self-evident axioms, and he builds a great system of geometry by pure reasoning.

The Greeks found this method so successful in geometry, that they tried to apply it to other subjects—what we should now call chemistry, physics and astronomy. But here they were not so successful. The difficulty lay in choosing the right truths from which to start, as we start from the axioms of geometry. It is very hard to discover these basic scientific truths. It requires methods of observation and experiment which the Greeks, for all their genius, never mastered.

6. *The Beginning of Science*

By the time of the Greeks, many of the useful arts, of which we spoke in Chapter 2, had reached a high level of development. And the craftsmen who practised these arts had come to know a great deal about the materials in which they worked. In all this knowledge of the craftsmen there was science of a kind. And it was one of the main sources out of which the modern sciences took their rise. The early chemists were to draw upon the experiences of the potter, the glass-blower and the metal-worker. The Egyptian surveyor measuring anew the fields uncovered by the Nile, used working rules which were some day to become theorems in geometry. And out of the priestly cult of the stars as divinities came the recognition of their orderly motions. But the sort of knowledge that we have come to mean by the word *science* was the creation of the Greeks. And it is worth trying to understand what science in this special sense really is before we embark on the story of one of the greatest scientific movements in history.

The finest scientific achievement of the Greeks lay in the sphere of geometry. That was because progress in that subject does not depend upon carrying out experiments, a procedure which the Greeks generally distrusted, and which they never properly understood. In geometry they were able to start from properties of space which are too familiar to need experiment; and then they argued logically from point to point, a procedure in which they excelled. The way in which the craftsman's knowledge

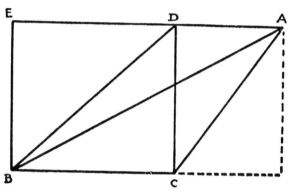

FIG. 2. THE AREA OF A TRIANGLE

is transformed into science is, in fact, best illustrated by means of a simple example from Greek geometry with which every reader will be familiar.

The Egyptian surveyors had a rule that the area of a triangle ABC is half that of a rectangle BCDE having the same base BC and the same vertical height DC (Fig. 2). The result of working by this rule must have been satisfactory in each particular case where it was used; but the Egyptians could not have proved that it must necessarily be exactly true for *any* triangle. This, however, the Greeks were able to do. For they had transformed, and vastly extended, the working rules of the Egyptian surveyors into a series of propositions such as we find today in any schoolbook on geometry. And one of these propositions proves to us that *any* two triangles on the same base BC and between the same parallel lines, BC, EA, are equal in area. Hence, in our figure, the triangle ABC is equal to the triangle DBC, which, in turn, is one half of the rectangle BCDE.

This example may also serve to illustrate one important feature of science—its *generality*. By this we mean that it is concerned with properties, or rules, which cover as wide a range of particular cases as possible, just as our rule was shown to hold good for *all* triangles. The scientist tries to discover *systems* of facts, and to fit these into other, wider systems. Our triangle falls into a system of triangles covered by the theorem, and the theorem occupies a place in a whole series of propositions forming a system of geometry.

Besides its generality and its system, there is one other feature that we notice about science, something that we express by saying that science is *disinterested*. By this we mean that the scientist is led to make his discoveries solely by his keen desire to know the truth about the world. He does not carry on research merely for the purpose of inventing useful "gadgets," or in the hope of making money out of them, though both these consequences may follow.

The career of Copernicus affords a striking example of this sort of thing. We shall see that, not only were his scientific achievements not valued by the people of his own time, but he suffered in reputation (and some of his followers suffered even more severely) for holding the opinions that he did. The motive power behind his thirty years of labor was simply the desire to understand the way the Universe worked.

This quality of disinterestedness in science is also something we inherit from the Greeks. Knowledge about nature in their day was simply the craftsman's knowledge, which was a collection of "dodges" for making and doing things. But the Greek thinkers who created science mostly belonged to a social class who were not interested in "dodges" for doing things. They were free men with many slaves to make and do whatever they wanted. There was not the same need to economize man-power that a modern manufacturer feels who pays wages to his work-people. And the thinker felt it no business of his to lighten the toil of the slave by inventing labor-saving devices. The ancient world cared less about human toil and suffering than we can well imagine.

This callousness offends our consciences today; but experience has proved that truth is best discovered by making it an end in itself, and not worrying for the moment about the good or bad uses to which it will be put when discovered. The Greeks did not study geometry just because they wanted to become surveyors. But out of their discoveries in geometry and astronomy there have come developments, even in surveying, beyond anything the Egyptians, or even the Greeks themselves, could have dreamed.

There were, however, other crafts which did not yield so readily to the sort of treatment that transformed surveying into geometry. These were the crafts roughly corresponding to what is now the science of chemistry. They were laborious crafts, involving the use of fire, and carried on by grimy, unlettered workers, from whom the Greek thinkers were almost completely cut off by their social position. Hence the chemical theory of the Greeks was a fiasco; that was the price they paid for neglecting the rich stores of knowledge about the stuff of the world which these humble craftsmen had accumulated through so many centuries.

On the other hand, the great rise of modern science in the last three hundred years dates from the time when men discovered that such knowledge must provide the raw material of science. The scientist has continually to go back to experience through the experiments he performs in his laboratory, in order to keep his thinking on the right lines. And as the experience of the craftsman served as the raw material of science, so today the scientist repays the debt by putting his knowledge at the service of the manufacturer. Purely scientific knowledge serves as the foundation of a whole range of modern industries, from the generation of electric power to the manufacture of silk stockings.

7. *Pythagoras of Samos*

All primitive peoples have picturesque stories to tell of how the gods made the world out of nothing, or out of the body of a giant or something of the sort. The Greeks had a whole collection of tales about the gods: Zeus, Apollo and the rest. They pictured them as glorified human beings living in style on the top of Mount Olympus.

But there were some Greeks who felt that such man-like gods could not possibly have made the Universe. They also felt that it was no good trying to describe the creation of the world by likening it to some familiar process, such as making a pudding. For you can only make a pudding in a world that is already a going concern. So these men gave up telling the old stories, and decided to look at the world with their own eyes.

As they looked out on the world, they saw new things coming into being and old things passing away. But things did not come from nothing, and they did not pass into nothing. There was something that went on existing all the time, but which was continually changing. What was that something?

This problem was first tackled by a group of Greek thinkers who lived in the sixth century B.C. on the coast of Asia Minor. They assumed that all things are composed of the same stuff, and they tried to discover what it was. One of them said it was water, which he thought could be condensed to form earth and solids, or expanded to form air. Another man said the ultimate stuff was air or vapor; another that it was fire, and so on.

But there was quite a different set of Greeks who thought that the important question was not What? but How Much? For example, we can make a pudding of flour, lard, eggs, sugar and what not. But the kind of pudding we get largely depends on *how much* of each of these we mix together; and *how long* we cook them. To take a less homely illustration: there are thousands of chemical substances all made of the three elements carbon, hydrogen and oxygen, yet no two of them are quite alike. Chemists tell us the difference between them is due to differences in the proportions and arrangements of the three kinds of atoms in the different substances. And if we turn from chemistry to other sciences, such as mechanics and physics, we find that the account they give of things also largely depends on counting and measuring and drawing diagrams, and upon other, more difficult, processes.

This whole idea that counting and measuring give us the clue to the understanding of nature goes back to a Greek called Pythagoras who lived in the sixth century B.C. The influence of his ideas played a great part in the story we have to tell.

Pythagoras seems to have been born on the island of Samos in the Aegean. He is said to have spent his earlier years traveling and studying in Egypt and the East. When he returned from his wanderings, he settled at Crotona in southern Italy. There he gathered round him a lot of young men who formed a sort of religious brotherhood something like a monastery. He proceeded to teach them by word of mouth instead of writing a book. It suited his fancy to keep his knowledge as the secret of his own brotherhood, and not to let it leak through to the vulgar herd outside. There was a story about a talkative Pythagorean who let out some secret information and who perished in a shipwreck as a judgment!

Pythagoras is said to have made simple experiments on the sound produced by plucking a stretched string— one of the very few reported instances of physical experiments among the ancients. He found how the pitch of the note given out depended on the length of string which was allowed to vibrate. This discovery convinced Pythagoras that simple arithmetical rules of this kind are the really important things in the Universe. He even thought that numbers were a sort of ultimate stuff out of which everything was made. He was also perhaps the first to set out geometrical propositions in a logical order as we find them in a geometry book today.

The views of Pythagoras colored his attitude to astronomy, and exerted a great influence in after years. He was almost certainly the first man to teach that the Earth is a sphere floating in space. But his reasons for believing this

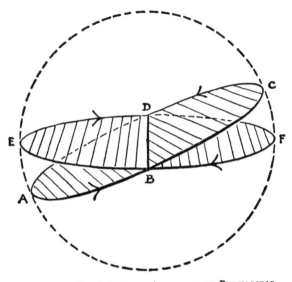

Fig. 3. The Sun's Motion According to Pythagoras

cannot have been the ones we find in a modern geography book. Probably he would have given as his main reason the fact that the sphere is the most perfectly regular of all solid figures.

Pythagoras, again, gave the simplest possible explanation of the way in which the Sun appears to move (Chap. 3). He thought it traveled right round the sky in a circle ABCD (Fig. 3) at a uniform rate, completing its circuit in one year. At the same time the whole sky revolved once a day in the direction of the circle FBED, carrying the Sun and its circle with it. This accounted for the Sun's daily rising and setting, and also for the way in which the length of the day, and the points on the horizon where the Sun rises and sets, vary according to the time of year. Pythagoras gave a precisely similar explanation of the Moon's apparent motion. He also tried to represent the motions of the five planets in the same way; but the attempt broke down because the behavior of the

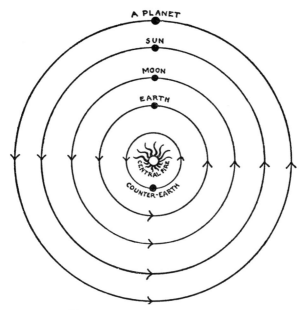

FIG. 4. THE SYSTEM OF PHILOLAUS

planets is more complicated than that of the Sun and Moon. However, the important point is that Pythagoras started the idea that the complicated movements of the heavenly bodies must be regarded as made up of a number of circular motions all put together. We shall see how this idea kept its hold on astronomy until about three hundred years ago.

The movement started by Pythagoras was carried on after his death. One of its later supporters called Philolaus (late fifth century B.C.) put forward the view that the center of the Universe was occupied, not by the Earth, but by a huge Central Fire (Fig. 4). Around this Fire the Earth revolved daily in a circle. He supposed that the Sun, Moon and planets all revolved, like the Earth, about the Fire. So did another mysterious body, the Counter-Earth, which, at least in the imagination of

Philolaus, revolved in an orbit between the Earth and the Fire. The effect of all this was to account, both for the daily risings and settings of the heavenly bodies, and also for the slow circulation of the Sun, Moon and planets against the background of stars.

We must not confuse the Central Fire with the Sun. Later writers did; and they regarded Philolaus as having invented the Copernican system some 2000 years before Copernicus. He did not do that; but he may have been the first man to think of the Earth as revolving round the center of the Universe together with the other planets.

8. The Philosophers

The Greek thinkers were at first mainly concerned with trying to understand the world around them. But later they became increasingly interested in people, and in discussing how men ought to think and behave. The man who did most to give this new turn to Greek thought was Socrates, who lived in Athens in the fifth century B.C. The movement he started was carried further by his fellow citizen and disciple Plato.

What these men sought above all to gain was not a mere knowledge of particular facts, but a right judgment as to what is true and good. The Greeks called this right judgment *sophia,* or "wisdom." They thought only God could be called wise. But men could be "lovers of wisdom" or *philo-sophers.* And, following them, we give the name of *philosophy* to the search after the most general truths concerning the world we live in, and after the kind of life we should lead.

The earlier thinkers, Pythagoras and the rest, who had tried to discover the permanent reality behind all changes (Chap. 7) had been philosophers too. But from Socrates onwards, philosophy branched off from the mere open-eyed study of nature to concern itself with these more general problems of truth and right living. With these developments we are not here directly concerned.

Philosophy continued, however, to exert an important influence upon the growth of the sciences of nature, and of astronomy in particular. And this influence continued to be felt throughout the whole period we shall cover in the following pages.

The influence of Plato himself on the progress of astronomy was not altogether favorable. We can best understand the new turn he gave to the subject by going back again for a moment to school geometry. This is all about straight lines, triangles, circles and so forth. But they must be *perfectly* straight lines, *perfectly* regular circles, etc. The diagrams in the book or on the blackboard, however carefully drawn, are not the perfect figures, but only crude imitations to help us to follow the argument. This is all right in geometry. But Plato and his followers thought that the same principle could be applied in the natural sciences, such as astronomy. They thought the heavenly bodies, and other objects we see around us, were not in themselves worth studying. They merely served, like the diagrams in geometry, to suggest to us ideal problems which must be solved by the reason and not through the senses. The result of this attitude was to discourage experiment and observation in Europe for nearly 2000 years.

Otherwise, Plato merely carried on the astronomical ideas of Pythagoras, ignoring the later developments which made the Earth move round the Central Fire. He pictured the Universe as a vast sphere having at its center the spherical Earth round which the Sun, Moon and planets revolved in concentric circles. The radii of these circles were supposed to stand to one another in a simple proportion which he could not measure, but merely imagined.

Like Pythagoras, Plato represented the motion of each of these seven bodies by a combination of two circular motions in nearly opposite directions (Fig. 3). He knew that this simple system broke down for the planets. So he set his pupils the problem of representing the behavior of the planets by means of still further uniform circular motions like the wheels within wheels of Ezekiel.

9. Aristotle

Plato had a pupil who became as great a philosopher as himself, though his ideas developed on different lines from those of his master. This was Aristotle, who lived from 384 to 322 B.C. His father was court physician in Macedon (Chap. 5) and he himself was tutor to the boy who grew up to be Alexander the Great. He studied under Plato in Athens, and in due course settled there himself as a public teacher.

Aristotle's teachings cover almost the whole field of knowledge. He corrected the tendency we have noted in Plato to treat the science of nature as if it were a kind of geometry. The parts of his work which deal with what we should call zoology are of great value. They are based on the actual observations of Aristotle and his friends, and show him to have been one of the greatest naturalists of all time. But in astronomy and physics, Aristotle was on less sure ground. He based his conclusions on superficial observations of how things appear to behave, or even upon his own ideas of how they ought to behave. However, Aristotle's mental picture of the Universe was widely accepted wherever it became known until right into the seventeenth century. It formed the backbone of the resistance which the ideas of Copernicus encountered. Hence we must try to give some account of it.

Aristotle pictured the Universe as a sphere of space of limited size. Everything that exists was somewhere inside this sphere; outside it there was nothing at all, not even empty space. At the center of the Universe was the Earth, a motionless sphere. Around the Earth, the Universe was built up in layers like a nest of boxes. The core of the Universe was of *earth*, forming the dry land. Over this was a layer of *water* forming the ocean; then the atmosphere of *air*, and then an outer coating of *fire* extending as far as the Moon. All the rest of the Universe was taken up by the spheres which carried round the Sun, Moon and planets. In the outermost sphere, the stars were fixed like silver studs.

The sphere which carried the Moon was like a wall

dividing the Universe into two regions. Inside it, everything was made of the four elements, earth, water, air and fire. These were supposed to be constantly changing one into another, so that we lived in a realm of change and decay. But on the other side of the Moon's sphere, the heavenly bodies and the spheres which carried them round were supposed to be made of a fifth element—the *aether*. (Do not confuse this with the ether which the chemist and doctor use today, or with the medium upon which radio waves are supposed to travel.) This aether was thought to be much superior to our four elements, for it was supposed to be incapable of any change except change of position, that is, motion. This idea of a sharp distinction between the corruptible region of the four elements and the incorruptible heavens took a very firm hold on people's minds. It was not completely abandoned until about a hundred years ago, when it was found that the Earth and the heavenly bodies are all made of the same sorts of matter.

Aristotle noticed that heavy bodies such as stones fell to the ground, while flames shot upward. He concluded that it was natural for the four elements to move in straight lines—earth and water downward towards the center of the Universe, and air and fire outward from the center. He supposed that a body might also move round and round in a circle, and that this would be a better kind of motion than moving in a straight line. For it could go on forever, while a body moving in a straight line would have to stop when it came to the boundary of the Universe or to its natural resting-place. Hence he thought that this better kind of motion must belong to the better kind of element—the unchangeable aether. Thus he proved to his own satisfaction that the heavenly bodies go round in circles while the Earth stands still in the middle. We find the argument very unconvincing. But it was about 2000 years before it began to be upset by sounder ideas on how bodies move; and in the meantime almost everybody believed that the Earth stands still in the middle of the Universe.

To simplify matters we have spoken as if each planet were carried round by a single sphere. But to represent the complicated motions of each planet required quite a number of revolving spheres, each giving its own motion to that planet. All this machinery was much too complicated. Moreover it took no account of the fact that the distance of a planet from the Earth obviously varies, for at some times it appears much brighter than at others. Hence a different method had to be invented to reduce the motions of the planets to a rule which would enable astronomers to predict their future behavior.

About the time that Aristotle died, the center of Greek scientific activity shifted from Greece to the city of Alexandria, founded on the Mediterranean coast of Egypt by Alexander the Great. It is with that city that the later developments of Greek astronomy are largely associated.

10. Greek Astronomy

There is a story that the old Indian astronomers pictured the Earth as a sort of flat tea tray supported on the backs of three elephants. The elephants, in turn, stood on the back of a huge tortoise; but what supported the tortoise we are not told! The Greeks started off by asking the same sort of questions about what the Earth is like and what keeps it in its place, what the heavenly bodies are made of, and so forth. But instead of talking about elephants and tortoises, they gave common-sense answers to these questions. They supposed that the Universe worked, on a large scale, by the natural processes which they saw at work, on a small scale, around them, and which they used in their crafts.

Thus one of the earliest Greek thinkers, Thales, in the sixth century B.C., said the Earth was a short cylinder (something rather like a pill-box) floating on a vast ocean of water; it had a flat top on which we live. Others said the Earth was a flat disc, floating like a leaf on the air, or that it was forever falling like a stone. Finally, Pythagoras, as we have seen, said the Earth was a sphere floating in space. This idea was generally accepted, although it was not until the early sixteenth century that Magellan's ships sailed round the world and proved that old Pythagoras was right.

The universal curiosity of the Greeks extended beyond the Earth to the heavens. Plato tells a story of the aforesaid Thales, how he fell into a well while star gazing, and was afterwards teased by a saucy servant girl for being so keen to know what went on in the heavens that he could not see what lay at his very feet. It was the Greeks who put the idea into people's heads that the sky is a globe of some glassy material in which the stars are fixed like silver studs. They learned that the Moon shines by sunlight, that it is eclipsed when the Earth comes between it and the Sun, and much else of interest.

However, the central problem upon which the Greek astronomers came more and more to concentrate their attention was that of accounting for the observed motions of the "planets" (among which the Greeks included the Sun and Moon as well as the five bodies that we call planets, see Chap. 3). It is with the story of efforts to solve this problem that we shall be mainly concerned in the following pages.

Any attempted solution of the problem takes the form of what is called a planetary *theory*. Such a theory can be represented by a diagram showing the path which a planet is supposed to follow in space, and specifying the rate at which it moves along this path. If such a theory is drawn up correctly, it will enable us to say where a planet will be at a given time, that is, to *predict* the future positions of the planet. It is possible then to draw up a table of such future positions, like a railway timetable which shows the times at which a train will call at the various stations.

However, there is always this difficulty about constructing planetary theories. We can see a planet moving against the background of the stars. But we cannot see it moving towards or away from us. Only the slight changes in the brightness of the planets, or in the apparent size of the Moon, prove that these bodies do move *in depth* and are not always at the same distances from us. So there was always a lot of guesswork in the construction of the early planetary theories. And even when they enabled tolerably reliable tables to be drawn up, no one could be sure whether the planets really followed the paths in space that they were supposed to in theory. However correct a railway timetable may be, it does not tell us anything about the route followed by the train between the stations. No more, for that matter, does it tell us anything about the mechanism of the engine which pulls the train along. And during almost the whole of our period, no one had the faintest idea *why* the planets move at all.

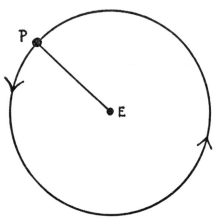

FIG. 5. THE SIMPLEST PLANETARY THEORY

Broadly speaking, the development of Greek planetary theories seems to have been as follows. The simplest possible planetary theory is to suppose that the planet P describes a circle at a steady speed about the Earth E considered as a mere point at the center of the circle (Fig. 5). This assumption fits the observed motion of the Sun fairly well, the motion of the Moon not so well, and the motions of the five planets not at all well. (We are speaking here about the motions of these bodies against the background of the stars; not about the daily motion of rising and setting which they all share with the stars, and of which we do not take account for the moment.) This, as we have seen, was the theory of Pythagoras. It was taken over by Plato, improved by his disciples, and then adopted by Aristotle (Chaps. 7-9).

But the later Greek astronomers rejected this theory because it made no allowances for the changes in the speeds of the heavenly bodies, or in their distances from us. They tried to alter the simple scheme of Fig. 5 in order to represent these changes. But in doing so they were restricted by the rule laid down by Aristotle that a heavenly body could only be thought of as describing

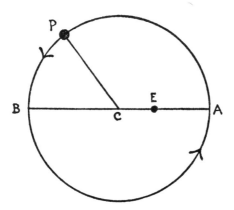

FIG. 6. THE EXCENTRIC

a circle, and that at a uniform speed (Chap. 9). So the Greeks tried to get over the difficulty in two ways.

On the one hand, they supposed the center C of the planet's circular path APB to be some little way from the Earth E (Fig. 6). When the planet P in its course reached A, it was nearer to the Earth, and it appeared to an earthly observer to be moving faster, than when it was at B. This device, called an *excentric,* served very well to represent the apparent motion of the Sun.

On the other hand, the Greeks sometimes supposed the planet P to describe a small circle called an *epicycle* about a center C, while C described a larger circle called a *deferent* about the Earth E at its center (Fig. 7). Here, again, at A, the planet was farther from the Earth and at B nearer to it. And when at B, the planet was moving the opposite way to C, and might appear to an observer at E (the Earth) to be moving back towards the west, as the five planets do appear to move when nearest to the Earth.

The Greeks combined these two devices, and developed them in ways that we need not go into. But what we have just said may give some idea of what a planetary theory was like in ancient times and through the Middle Ages right down to the time of Copernicus. The relative sizes of these various circles, and the speeds of the planets therein, were inferred from the observed positions of the planets at known times.

While thinking along these lines, the Greeks actually hit on the idea that the Sun might be the fixed center about which the Earth and the planets moved in circles. This interesting development began in the fourth century B.C. with one Heraclides trying to account for the peculiar behavior of Venus and Mercury. These planets are never seen far from the Sun, and they appear sometimes on one side of him and sometimes on the other. Heraclides suggested that perhaps they each described a circle about the Sun, while he revolved about the Earth, just as, in Fig. 7, P describes a circle about C while C revolves about E. It was natural to extend this to the other three planets, Mars, Jupiter and Saturn.

Then it was realized that everything would look just the same if, instead of the Sun revolving round the Earth, the Earth went round the Sun, just as the five planets were being supposed to do. This step was taken in the third century B.C. by a Greek astronomer called Aristarchus who came from the same island of Samos as did Pythagoras. He also accounted (as Heraclides too had done) for the daily rising and setting of the heavenly bodies by supposing the Earth to turn round once a day on its axis. He thus arrived at the complete Copernican theory of the solar system, and earned his title of the "Copernicus of Antiquity."

Aristarchus did not work out his theory in detail as

Copernicus did. He did not make use of it to calculate planetary tables as other Greek astronomers soon did from theories which made the Earth the fixed center of the Universe. And in any case the idea of a moving Earth was very little to the taste of people brought up on Aristotle's philosophy of nature. So little more was heard of this "Sun-centered" planetary system until 1800 years later, when Copernicus began to establish it as the accepted theory of modern times.

In the meantime, an "Earth-centered" planetary theory had been worked out in great detail by the later

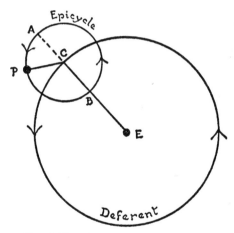

FIG. 7. THE EPICYCLE AND DEFERENT

Greek astronomers—notably by Hipparchus of Rhodes (second century B.C.) and Ptolemy of Alexandria (second century A.D.). It was made the basis of tables which could predict the positions of the planets to a degree of accuracy not much less than that to which their positions could be determined with the instruments of the time. These instruments underwent little improvement until after the death of Copernicus. Some idea of their nature may be obtained from our account of his attempts to make similar ones (Chap. 21 below). It is important to bear in mind that no telescopes were in use until the early seventeenth century.

All of Greek astronomy that succeeded in establishing itself was collected into a famous book by the aforesaid Ptolemy, the last of the great Greek astronomers, somewhere about 140 A.D. The book has a Greek title meaning the *Collection;* but it came in the Middle Ages to be generally called the *Almagest*—the word is a mixture of Greek and Arabic. Ptolemy's *Almagest* and Copernicus's *Revolutions* (Chap. 29 below) have been fancifully called the Old and the New Testaments of Astronomy. They are separated by some fourteen centuries, during which the "Earth-centered" picture of the Universe (Fig. 14A), as conceived by Aristotle and Ptolemy, was everywhere accepted.

11. The Centuries Between

Alexandria, where Ptolemy lived, was, in his time, part of the Roman Empire. The Romans made practically no contribution to astronomy; but they come into our story for other reasons. The Romans were originally a small group of tribes living in the part of Italy where Rome stands today. But, by the time of Christ, centuries of conquest had put them in control of all the countries bordering upon the Mediterranean, as well as of what is now France and much of Britain. In the long years of Roman peace which followed, Latin, the Roman tongue, became a universal language in western Europe. The Christian Gospel spread from Palestine all over the Roman world in ways with which we are familiar from the *Acts of the Apostles.* And early in the fourth century, the whole Roman Empire became nominally Christian.

The Greeks had long since lost their vigor, and, as time passed, the great mass of Roman citizens began to lose interest in public affairs; their population also declined. In consequence, the Germans and other vigorous peoples of Europe were able, from the end of the fourth century A.D. to pour into the Roman Empire and to settle on its land. These barbarian invaders were anxious to preserve Roman civilization and to learn from it. They became Christians; and it was the task of the Church to pass on to the barbarians whatever knowledge had survived from the ancient world.

Very little knowledge had survived in the West. Hence we get a period called the Dark Ages, extending roughly from 400 to 1000 A.D., during which very crude ideas prevailed about the world of nature. Some people even denied that the Earth was a sphere; or, they said, if it were, no one could live on the southern part of it without falling off. Others thought the whole Universe was arranged like the Tabernacle of Moses, described in the Old Testament. Later, some books were discovered which set forth Plato's ideas about what the world was like. These were brought into line with Christian teaching, and they furnished the peoples of western Europe with a sort of mental picture of the Universe which lasted without much alteration down to the sixteenth century.

People thought of the Universe as an enclosed space —a sphere of strictly limited size. It had been created by God in a not very distant past as the abode of man, for whom the whole world of nature existed. It was destined to pass away, or to undergo some catastrophic change, in the not very remote future. The Earth occupied the center of the Universe, both in position and in importance. It was the stage on which the drama of human life was played out. Under our feet, in the Earth's interior, was Hell. Outside the boundary of the Universe was Heaven, the abode of God and the Saints. Since the whole Universe was thought to form a single system, it was natural to go on believing that the stars in their courses controlled or foreshadowed events on the Earth. There was also thought to be close correspondence between the various star-groups and the parts of the human body.

While this Dark Age overshadowed Europe, a new religion had sprung up among the Arabs living on the far side of the Red Sea. It was founded in the seventh century by Mohammed, who proclaimed himself a prophet of God. Under the inspiration of their new faith, the Arabs swept forth in a great wave of conquest which carried them in the East to India, and in the West through Egypt and North Africa to what is now Spain. When the force of this wave had spent itself, the Arabs began to settle down and study the great writings of the ancient world, which they translated into Arabic. Among these works was Ptolemy's *Almagest* (Chap. 10) and the scientific works of Aristotle. A great revival of ancient science thus took place among the Moslems, or followers of Mohammed, but little further progress was made by them.

In Spain there was a frontier between Christendom and Islam (the Moslem power), and in Sicily there was another. Across these frontiers, from about 1000 A.D. onwards, the ancient learning flowed from the Moslems to the Christians. Many a thrilling story could be told of scholars from France or England who lived for years among the Moslems, often in secrecy and in peril of their lives, learning the difficult language and acquiring the precious manuscripts. In this way, sometime in the twelfth century, the Christian scholars of western Europe gained possession of Ptolemy's *Almagest* and Aristotle's scientific works, not yet in the Greek originals, but in Latin translations from the Arabic.

In the West, as we have seen, the lamp of the old learning had never quite gone out. It was kept dimly alight, first in the Benedictine monasteries which sprang up all over Europe in the Dark Ages, and later at the schools which grew up round the great cathedrals.

These schools were chiefly designed to turn out young priests who could read the Bible and whatever other books they needed to carry out their duties. But they were also taught a smattering of astronomy, at least enough to understand the rules for fixing the dates of Easter and other Church festivals. Another thing the young priest was taught was the elements of logic—the science of right reasoning—as laid down by Aristotle, and accepted as obviously true. As a writer on logic, Aristotle had long been known in the West. And now in the twelfth century, when his writings on a wide range of other subjects were being recovered, people naturally thought that his books on such subjects as physics and astronomy must be as true as his other works, even as true as logic itself. But they were wrong. For Aristotle's views on the physical sciences were hopelessly astray.

However, the young priest was encouraged to use Aristotle's logic as Aristotle himself had used it, to argue out all sorts of interesting conclusions, first in theology and philosophy, and, later on, in the sciences of nature. In this way there was built up in the cathedral schools and the rising universities a vast system of belief about God, man and the world. It came to be known as the philosophy of the schools, or the *scholastic* philosophy. It represented an outlook which, in one form or another, held its ground until roughly the end of the sixteenth century.

On the one hand, then, this system took over from Aristotle the belief that the Earth was the center of the Universe. On the other hand, it was brought into line with Christian theology, so as to be a sort of official philosophy of the Church. That was how it came about that Copernicus and his followers, who taught that the Earth moved round the Sun, found themselves in conflict with the religious authorities of their age.

12. The Age of Copernicus

Copernicus lived at a time when vast and complex changes were taking place in Christendom. Powerful forces were at work which influenced both the course of his own life and the progress of his theory in the years following his death. It may help us to understand the astronomer's life story if at this point we take a brief glance at the European scene round about the beginning of the sixteenth century.

Hitherto Europe had been mostly divided up like a chessboard into little states. The inhabitants, from the prince down to the peasant, were mostly linked together in a chain of rights and duties. Here and there, however, were towns and cities which had freed themselves from this chain and which flourished by making and selling things, or by trading the natural products of the country, often overseas. Many of the peasants in central and eastern Europe wanted to go to the towns to make more money in this way. But the nobles among whom the land was portioned out would not let the peasants go. They kept them to toil in the fields for next to nothing, while they themselves waged little wars on each other, or on the rich towns. In order to protect their merchandise by land and sea from pillage and piracy, the trading towns banded themselves together. One of the oldest and most extensive of such alliances, the Hanseatic League, included some eighty towns of northern Europe, with several of which Copernicus was closely connected.

While Europe was split up in this way politically, the Church formed a sort of superstate covering the whole of western Christendom. It was ruled from Rome by the Pope (the Bishop of Rome) and his cardinals. Europe was divided up into dioceses (under bishops) and these in turn into parishes (under priests). In every land the clergy formed a separate class exempt from the law to which the other inhabitants had to submit. The Church everywhere controlled education, so that no one could teach anything of which it did not approve. Those who tried to do so were speedily silenced. It also touched the life of the individual through its ministries in baptism, marriage, burial, the bequest of property to heirs, and so forth. The fees for these services, together with the tithe, or Church tax of a tenth part of the produce of the soil, represented a steady drain on the resources of Christendom. This was increased by the growing practice of offering pardon for sins in exchange for a money payment. Also much of the land had been getting into the hands of the Church or of its orders of monks.

Worst of all, the Church had lost her spiritual vocation. Many of the priests were selfish and lazy and grossly ignorant of what it was their business to teach. Much of what they taught had nothing to do with real Christianity, as, for example, the sale of pardons just mentioned. They did not go to the Bible for their religion, but took it at second or third hand, from commentaries or traditions in which all sorts of unscriptural elements had entered. The Popes were sometimes men of exceptionally evil life. They had a little state of their own, and they played at power politics, backing one side or the other in the wars of the age as best suited their own interests.

In the days of Copernicus, this whole state of things was breaking down. Europe was forming into large states, roughly corresponding to those into which it is divided today. The little nobles were giving place to princes or strong central governments; and this made for prosperity, for now the workers were freer to go where they were needed, and the little private wars, which interfered with trade, were stopped.

Moreover, large parts of Christendom, notably England and northern Europe, were breaking away from the control of Rome in religious matters. This movement had a political and a spiritual side. Politically, it was an attempt to help trade by stopping the flow of wealth to Rome; also to unite and strengthen each country by bringing the priests under the same laws as everyone else. Spiritually, it was an attempt to get back behind the Middle Ages to New Testament Christianity.

This return to a purer faith was itself part of a larger movement, of which the return of Copernicus to a more primitive astronomical theory may perhaps be regarded as another aspect. This larger movement, generally called the Renaissance, was a great mental and spiritual experience through which Europe passed at this period. It marked the change-over from the Middle Ages to the modern age in which we live. It was a complex and long-drawn process, taking different forms in different countries.

From one point of view, it was a "Revival of Learning," particularly of Greek learning, resulting from the transfer to Italy of what was left of the old Greek civilization. When the western, or Latin, half of the Roman Empire went down before the irruption of the Barbarians, the eastern, or Greek, half, with its capital at

Constantinople, had managed to survive. But for some time now the Moslems, long since checked in the West, had been pouring into Europe from a new angle. In 1453, twenty years before Copernicus was born, the Turks captured Constantinople. Anticipating this disaster, the Greek scholars had for some time been streaming westward to Italy, bringing with them the manuscripts which enshrined the old literature and science. Italy thus became the fountainhead of what was called the New Learning, which was to exert a potent influence upon Copernicus during his ten years' sojourn beneath the Italian sky.

With the recovery of Greek learning, men began to have a clearer vision of the past. But, what was more important, they caught something of the spirit of the old Greeks themselves. They began to think for themselves, and to question beliefs that they had hitherto accepted on the authority of old books. There was also a new mood of enthusiasm and vitality, and an impulse to push into every field of experience, good or bad. This was partly the motive and partly the result of the spectacular discoveries made by the explorers of this period. Copernicus was nineteen years old when Columbus discovered America.

The Revival of Learning notably influenced astronomy in the fifteenth century. This was partly through the recovery of ancient astronomical books in the original Greek, and partly because, in the spirit of the time, men started to observe the heavens again for themselves. The science was stimulated by the demands of sailors for the nautical instruments and tables they needed for traveling on their new ocean highways. And, like all other branches of learning, astronomy profited by the invention of printing, which, in the lifetime of Copernicus, generally superseded the tedious and faulty copying-out of manuscripts as a means of preserving old books and of spreading new ideas abroad.

An outstanding representative of these new tendencies in fifteenth century astronomy was John Müller, generally known by his Latin name of Regiomontanus ("the man from Königsberg"). He made observations and discovered errors in the accepted planetary tables of the period. He traveled to Italy to learn Greek, to study Ptolemy's *Almagest* in its original tongue, and to translate it into Latin. He finally settled in Nuremberg, where he found a wealthy patron, Bernhard Walther, and craftsmen who made astronomical instruments for him and printed his books. From his press he issued almanacs and other aids to navigation which were used by several of the famous mariners of the period.

With Regiomontanus we reach the end of this introductory sketch; for when he died, in 1476, at the early age of forty, Copernicus was three years old.

PART TWO

The Man and His Work

13. Parentage and Birth

Whenever anything new and startling happens in the world around us, we instinctively try to find out the cause of it. This same instinct prompts us to seek an explanation of the mysterious fact of scientific or artistic genius when it confronts us in a Copernicus, a Shakespeare, or a Menuhin. One kind of explanation consists in unearthing a line of distinguished ancestors from whom the genius may, in some degree, have inherited his gifts. Hence in writing the life story of a great man, we usually begin by trying to trace his descent from earlier generations.

We know scarcely anything about the ancestors of Copernicus. And even the story of his life has had to be pieced together from all sorts of scraps of information which have somehow survived. There are still many gaps in our knowledge. The people who lived all round the astronomer could have found out and written down much that would have been of great interest to us today. But because of their indifference, or because of the obscurity in which the astronomer lived and died, his contemporaries allowed these precious memories to fade. It was not until Copernicus had been in his grave a hundred years and more that the first reasonably good biography of him was written. And that was the work of a Frenchman, Gassendi, who was hopelessly out of touch with the circles in which the astronomer had lived and labored. Hence most of what we know today about our subject has been laboriously dug out from old records by Copernican students, some Polish, some German, during the last hundred years.

The first problem we have to face arises out of the fact that, in the days when Copernicus lived, no one worried much about the spelling of names. For instance Shakespeare, who lived some time later, spelled his name in all sorts of ways. Hence it is a matter of some uncertainty what precisely the astronomer's family name was. Perhaps we may take *Koppernigk* as the most usual spelling. In any case, he soon abandoned the name to adopt the Latin version of it by which he is always remembered.

We first come across the name of the Koppernigks as a place-name in Upper Silesia, on the eastern side of Germany. It is spelled in various ways in Latin documents of the thirteenth century. Probably it had something to do with the local copper-mining industry in which so many of the inhabitants of those parts were employed. Now people often used to be called after the places where they lived; and, by the fourteenth century, Koppernigk had become a family name. At that period quite a number of Silesians were moving into the neighboring country of Poland, which had once contained Silesia within its frontiers.

The Poles play a considerable part in our story. They first emerge into history as a group of tribes of Slavic speech, struggling for existence against the enemies who ringed them round. In the tenth century they adopted Christianity and became a European power. A great age of Polish history began in 1386. In that year, the reigning Polish princess, Jadviga, married Jagiello, the Grand Prince of Lithuania, the rich land to the east of Poland. Thus Lithuania was united to Poland under the one royal house of the Jagellons. With her eastern flank secured by the forests and marshes of Lithuania, Poland was able successfully to resist the Germans who had been thrusting eastward along the Baltic coast. From time to time, members of the Jagellon family reigned over the neighboring countries of Hungary and Bohemia, so that the influence of Poland extended all over eastern Europe.

Like America in the early days, Poland was a big undeveloped country, only too glad to admit people from all quarters. Hence many traders and adventurers of all kinds from what is now Germany went flocking into Poland. There were also persecuted Jews and, later, Protestants. These all helped to form a vigorous middle class which filled the social gap between the Polish nobility and the peasants.

Among those who migrated from Silesia into Poland in the fourteenth century, there were people called Koppernigk who, as the records show, settled in Cracow, Torun and other Polish centers, where certain of them acquired rights of citizenship. Some of these people may have been among the ancestors of the great astronomer; but we know nothing for certain.

The first of the family to emerge into the light of history is one Nicholas Koppernigk, the astronomer's father. We know that he was a merchant, that he lived for a time at Cracow, and that, not later than 1458, he moved to Torun where he died in 1483. He must have been a man of character and business ability, for he became one of the leading citizens and, indeed, a magistrate of Torun. There he married into the wealthy family of the Waczenrodes, or Watzelrodes. They, too, were of Silesian origin, but they had lived for generations in Torun, and there, from time to time, members of the family had held office as councillors or magistrates. It was to a daughter of this house, Barbara Waczenrode, that Nicholas Koppernigk was married some time not later than 1464.

There were four children of the marriage—two sons and two daughters. The girls hardly come into our story. The elder sister, Barbara, became a nun, and, in due course, an Abbess, at Kulm. The younger, Katherina, married a merchant of Cracow. The future astronomer was the youngest of the family. He was born on February 19, 1473, in a house in one of the streets of Torun, formerly St. Anne's Lane, which is still pointed out as his birthplace. He was named Nicholas after his father. His elder brother, Andrew, we shall meet again as the companion of Nicholas in his travels and studies.

14. Boyhood

We know nothing about the home life of the young Nicholas Koppernigk, but we can form a fairly clear idea of the sort of world in which he grew up. And it is the familiar scenes, the little circle of relatives and friends among whom the child's earliest years are spent, that make the deepest impression upon him, rather than the lessons learned later at school and college.

The Greeks had a saying that "the city teaches the man." We today are organized into great national states numbering millions of inhabitants. Our daily lives are regulated by central governmen.s in which we feel only a remote interest. But in fifteenth-century Europe it was different. There, as once in Greece, the city was the vital and almost self-contained unit. The life of such a city, ordered with dignity and picturesque pageantry, must indeed have been an education to those who grew up in its midst. Particularly must this be so when the child is a future Copernicus, when his city is a perilous frontier-town, and when his father and his uncles are among the leaders of the little city-state, helping to shape the course of history in stirring times.

The town of Torun had been founded on the Vistula in the thirteenth century by the Teutonic Knights, a German order of military monks. They were rather like the Knights Templar, who fought in Palestine to drive the Moslems out of the Holy City of Jerusalem. The Teutonic Knights, however, were more interested in European heathens. Now that the Crusades were coming to an end, they decided to carve out a new country for themselves on the east side of Germany by conquering and converting the pagan Prussians. They set up Torun as a sort of fortified outpost, marking the boundary between their territory and that of the Poles, who lived to the south and east. The Poles had been harassed by the Prussians, and they welcomed the arrival of the Knights, but they soon found that they had made a change for the worse.

At the time our story begins, the Knights had been having trouble with their Prussian subjects who had risen in revolt against them. There had been a war in which Poland had helped the Prussians. The Knights had lost the western half of their little state, and they held the eastern part only as vassals of the Poles. Thus, less than ten years before the astronomer's birth, the town of Torun had come under the protection of the King of Poland.

Although Poland lay on the outer fringe of western Christendom, it was one of the most civilized of the states of Europe, at least as far as its city life was concerned. Torun had once been a prosperous center of trade between western Europe on the one hand and Poland and Hungary on the other. By the time Nicholas was born, however, most of this trade had passed to the great port of Danzig at the mouth of the Vistula. Still he must often have watched the ships sailing up the great river bringing manufactured goods from Germany and Flanders to exchange for cargoes of minerals from the mountains of central Europe. It was there, perhaps, in the business atmosphere that hung about his father's warehouse, that the boy acquired the knowledge of affairs on which he was to draw in later years when he was called upon to reorganize his country's coinage.

As a relief from business cares, there would be excursions on summer evenings to the family vineyards outside the town by the riverside. Nor would the talk be only of trade. The Magistrate and his family could mix freely with people of the highest social standing. Many of them were patrons of art, music and letters, and the boy would early be brought into touch with these things.

Learning was represented in the family circle by the mother's brother, Lucas Waczenrode, a Catholic priest and a distinguished scholar.

However, the little circle was destined all too soon to be broken up. When the boy Nicholas was only ten years of age, his father died. All the four children were then adopted by this priest, their maternal uncle. He was a second father to them; and he played such an important part in the astronomer's career that a word must now be said about him.

Lucas Waczenrode was born at Torun on November 29, 1447. At the age of sixteen he went to Cracow University. Later, as a young man, he crossed the Alps to study in Italy, where he won high honors, graduating at Bologna as Doctor of Canon Law (Church Law) in 1473. After a spell as a schoolmaster in his native town, Lucas entered the service of the Church, where he won rapid promotion. By 1479, he was a Canon of Frauenburg Cathedral, in Prussia. Later he went on a mission to Rome. In 1489 (six years after adopting the orphan family) Lucas was consecrated Bishop of Ermland or Warmia, one of the four dioceses into which Prussia had been divided for purposes of Church government. Ermland formed something like a little state (a vassal state of Poland) of which the Bishop was the ruler as well as the chief of the clergy. The Bishop's Palace was at Heilsberg; his Cathedral church was at Frauenburg on the coast (see Fig. 9 on p. 72).

Bishop Waczenrode seems first to have sent his nephew to the St. John's School at Torun where he himself had been a master. Later the boy attended the Cathedral School of Wloclawek, some way up the Vistula from Torun. The course there was intended as a preparation for entrance to Cracow University. There is a story about a master at this school whose name was Wodka (meaning brandy) but who latinized it as Abstemius (meaning an abstainer). He was an authority on constructing dials; and the story goes that he and Nicholas Koppernigk between them constructed the sundial on the south wall of Wloclawek Cathedral.

15. *Cracow*

When Nicholas was eighteen years old, he said farewell to his friends at Torun and set out for Cracow, the Polish capital, to study at the great University there. His uncle the Bishop may already have decided that he should enter the service of the Church; and this was the first step to be taken.

Promising young men from Torun generally went to Cracow, Leipzig or Prague University. At that time Prague had fallen out of favor with strict Catholics because John Hus had started a movement there which had developed into a form of Protestantism. The choice naturally fell upon Cracow, where Nicholas would feel less isolated than at German Leipzig. His married elder sister would presumably be living there. It was his father's old home town, and there would be business connections with the place. Moreover, Bishop Waczenrode was himself an old student of Cracow, and he could give his nephew many valuable hints about the conditions in the University. The highest social circles would be open to a student whose uncle was the ruler of Ermland and in close touch with the Polish court.

Cracow was then a city famous throughout Europe for its wealth and culture. It formed a great trading center, bestriding the road that joined central and western Europe to the East. In its palace dwelt King Casimir IV, a member of the great Jagellon family, which had raised Poland to the proud position she then occupied.

The Cracow University had grown out of a school founded in 1364 by an earlier Polish king, Casimir the Great. It had been re-established by Jagiello in 1400, and it had since flourished under the patronage of the kings of Poland. The University taught all the usual subjects of study. But it was one of the earliest among the northern academies to be influenced by the new spirit which had arisen in Italy through the rediscovery of Greek literature. That spirit prompted men to study the ancient books for the sake of the light they could throw on the

world of nature and man. It bade them give up the dry
logical arguments which the Middle Ages had regarded
as the only way of discovering truth. Nicholas began to
be influenced by this spirit of "humanism" at Cracow,
and he was to drink more deeply of it in his later years in
Italy.

At Cracow, then, some time in the winter of 1491-92
the name of "Nicholas the son of Nicholas of Torun"
was entered in a register which is still preserved, along
with those of about seventy other new students. The sur-
name of Nicholas is not given; but there is a note to the
effect that he has paid his entrance fee in full. Four other
young men from Torun matriculated on the same occasion, among them an "Andrew" who may well have been
the astronomer's elder brother. The two young men may
have lived in a students' hostel, as most undergraduates
were required to do. Or they may have had rooms in their
married sister's house.

At Cracow University there were many foreign students. They came from Germany and Hungary, and even
from Italy, Switzerland and Sweden. They conversed together in Latin, the universal tongue which some people
would even now like to see adopted as a world language,
like Esperanto. It was at Cracow that Nicholas acquired
that mastery of the Latin language which he shows in all
his writings. A knowledge of Greek, however, was still
rare in that part of the world, and the language was not
yet taught at Cracow. In speaking Latin, it was natural
and convenient to turn the names of places and people
into words of that language. Scholars accordingly coined
for themselves Latin (or sometimes Greek) names. It
may well have been at this stage of his career, as he joined
the polyglot throng of students, that Nicholas Koppernigk exchanged his Polish name for the Latin form of it
by which he has been immortalized. However that may
be, this will be an appropriate point in our story to take
that step, and so we shall henceforward write of Copernicus the astronomer.

It was the common practice in a medieval university
for a freshman to begin by attending courses in the Fac-

ulty of Arts, whatever profession he intended to take up, whether divinity, law or medicine. Copernicus followed this rule; though it must be remembered that an arts course in those days contained many things that we should now class as science. In fact the practice of separating arts and sciences, and of giving degrees in science, pure or applied, is quite a recent development, adopted within the memory of people still living. The lectures at Cracow would mostly consist in reading and explaining the Latin classics and the great books which the University regarded as authorities on the subjects with which they dealt. We know that the young Copernicus attended some half-dozen courses of lectures on such standard textbooks of the day. These included Euclid, and they covered philosophy, astronomy, astrology, geometry and geography. We even know the names of the distinguished scholars who gave these lectures. There would also be disputations, or formal debates on set subjects, between the professors or the senior students. The matters debated were decided, not on their merits, or according to the evidence, but by appeals to the opinions of old writers.

The teachers in the Polish University would not all be Poles. They would be drawn from any or all of the Christian countries, notably from Italy. And when later Copernicus went to Italy, he would find many Poles teaching there. For once a man had become a Master of Arts at any recognized university, he had the right to teach at any other.

One of the greatest men at Cracow in those days was the Polish mathematician Albert Brudzewski. For some reason Copernicus does not appear to have attended his lectures. But he was personally known to Brudzewski, and was influenced by him in various ways. It was doubtless at Cracow, and under the inspiration of Brudzewski, that Copernicus was won over to the pursuit of astronomy. Here he laid the foundations of his knowledge of the subject, and learned how to handle astronomical instruments and to make observations of the heavens. However, judging from the works of Brudzewski that have come down to us, he did not borrow his revolutionary ideas at any rate from the *public* teachings and writings of this master.

There was a flourishing school of astronomy at Cracow, though the subject was still taught from the medieval point of view. That is, the movements of the heavenly bodies were still explained by reference to Aristotle's system of physics. The subject was chiefly studied as a means of keeping the Church calendar in order. It was also mixed up with astrology—the old Babylonian superstition of trying to foretell future events from the positions of the heavenly bodies (Chap. 4). Another motive for the study of astronomy which was just becoming important was its application to ocean navigation. Sailors had given up merely coasting along the shores, and had taken to sailing the high seas. They used simple astronomical instruments and specially calculated tables in order to try and find out from day to day where in the world they were. It was while Copernicus was at Cracow that Columbus discovered America.

During his years at Cracow, Copernicus began to make a collection of books on mathematics and astronomy, some of which he kept as long as he lived. Many of these books have been preserved to our own day. Some of them contain jottings and calculations in the handwriting of Copernicus which date right back to the days when he was a student at Cracow. Some of these calculations give us the impression that he had begun to take the first steps in the construction of his new system of astronomy even at that early stage in his career. However that may be, by the beginning of 1496, perhaps earlier, he was back at Torun, his years of study at Cracow ended.

When he left the University, it was entering upon a rather stormy period in its history. King Casimir had died in 1492. Copernicus would see the funeral, and would meet his uncle when the latter came up to Cracow with the nobility to help choose the new king. From the frontiers came war-scares about the Turks and Tartars. And in the University itself, not all the students took kindly to the new tendencies in education; they split up into rival gangs. Generally the cleavage followed national lines, producing the spectacle of German "humanists" fighting Hungarian "scholastics" in the streets.

We do not know whether Copernicus took any degree at Cracow. Very possibly he did not. Of the seventy or so Arts men of his year, barely a fifth proceeded to the B.A., and only one to the M.A. But he always had happy memories of his days in the Polish capital. He corresponded with former teachers and fellow students, and the official duties of his later years afforded him welcome excuses to revisit the city time and again.

Bishop Waczenrode was anxious to see his nephew properly settled in life, with enough money to live on. One way of effecting this was to have him appointed to some office in the Church which would bring him in a steady income. He reckoned on being able to use his influence to have his nephew made a canon in his own Cathedral at Frauenburg. True, Copernicus was only 22; but that need be no objection. In those days there was a case of a boy of 14 who was made a cardinal. The number of canons was limited; so it was necessary to wait for a vacancy to occur in the Chapter—the committee of clergy who helped the Bishop to run the Cathedral and diocese. One of the canons conveniently died. But that month it was the Pope's turn to choose a successor, and so the first attempt to secure the election of Copernicus to the Chapter came to nothing.

Nowadays we should think it wrong for a man to become a priest or a minister for the sake of making a living. But in those days very few people thought anything about it. Some people, however, did object to this and to even worse scandals in the Church. Their protest was part of the great religious and political movement to which we referred in Chapter 12. On the one hand there was the Reformation which led to the Protestants' breaking off from the main Catholic Church to secure stricter discipline and more Scriptural beliefs. On the other hand there was the Counter-Reformation by which those Christians who stayed in the Catholic Church raised the spiritual standards of its priests to the vastly higher level which we almost everywhere find today. Copernicus was a

man in middle life when Luther broke with Rome. The
echoes of the great controversy, quickly reaching the eastern limits of Christendom, were to disturb the astronomer's last years.

16. Bologna and Rome

While awaiting the next vacancy in the Frauenburg
Chapter, Bishop Waczenrode decided that his nephew
should continue his education, this time in Italy. If he
was going to be a Churchman, he had better learn something about Canon Law—the legal code, drawn up by
Popes and Councils, according to which the Church gov-
erns itself.

The most famous school of law in Italy was the one at
Bologna, which dated from the beginning of the twelfth
century. Accordingly it was to Bologna that Copernicus
repaired in the first instance, setting out from Ermland,
where he had been staying with his uncle, in the autumn
of 1496. Students from Prussia bound for Italy had the
choice of two alternative routes. One led through Vienna
and over the Semmering Pass to Venice. The other went
through central Germany to Augsburg, and thence over
the Brenner Pass to Verona. Copernicus seems to have
chosen the latter route, calling at some of the famous
German cities on the way, notably at Nuremberg, where
Bernhard Walther, the instrument-maker and patron of
astronomers, still lived.

In those days Italy was the fountainhead of the newly
recovered culture of Greece and Rome. Students used to
trek thither across the Alps from northern Europe. Under the blue Italian sky they learned Greek from the lips
of native teachers who alone understood the intricacies of
the old language. Or they explored the Latin literature
amid the magnificent monuments and the haunted land-
scapes which had seen its birth. Then there were the
pilgrims who were drawn to pontifical Rome by the aura
which still clung around the throne of the Vicar of Christ

and the graves of the martyrs. Lucas Waczenrode had happy memories of his own student days beyond the Alps. But Italy was no longer what she had been. Copernicus found the country ravaged by the French invaders, and sinking into an appalling state of wickedness. Rodrigo Borgia, a secret poisoner, was Pope, while the godly Savonarola, fearlessly denouncing the evils of the time, was qualifying for the scaffold.

Bologna was one of the oldest of the European universities. It had come into being almost by accident when the fame of a great teacher had brought students flocking to his feet from all over Europe. In the Middle Ages, a man who left the place where he belonged and went to live in some strange city found himself at a great disadvantage. He had none of the rights which belonged to the regular citizens. It was thus natural that university students should band themselves together for mutual protection and should demand recognition from the citizens. It was also natural that a city should grant privileges to the students in return for the increased prosperity which their presence brought to the trade of the city, and for fear lest they should migrate elsewhere and take their custom with them, as they often did. Hence the student community in a city came to form almost a little state with its own rulers and laws.

The teachers tended to form another group bound together for mutual support. It was their special concern that no man should enter their ranks or share their privileges unless he first proved his fitness to do so. Thus examinations and degrees began as the recognized avenue through which a scholar could pass from the student ranks to the privileged class of the teachers, or *Doctors*. A student taking a degree was like an apprentice becoming a properly qualified craftsman in one of the craft-guilds of the time. In fact the word we translate as *university* was often used of the body of skilled craftsmen who made up some particular guild. But it came in time to mean more especially the body of students, or the body of teachers, assembled at some particular place, or the whole institution in which they played their parts.

In some universities, as at Paris, the association formed

by the teachers was all important; and that is the system with which we are most familiar today. But Bologna University was run by the students themselves. It was governed by a Rector elected every two years. He acted on the advice of a council; but in the last resort his authority was derived from the mass meeting of all the students. The Rector's office was no bed of roses, and he often accepted election under pressure, as the Speaker of the House of Commons traditionally does. His troubles began with his election. At one time it was the custom for the students to celebrate this event by tearing the clothes off the new Rector's back, and then selling him the pieces at a ruinous price.

Student life at Bologna was fairly free and easy. It was the professors who were kept in order by strict regulations! A teacher had to swear to obey the Rector, or be deprived of the right to teach at all. He was fined if he fell behind schedule in his course of lectures, or if he left anything out, or if he was late in starting his lecture or went on speaking too long. He was also fined if the attendance at his class fell below a certain number. He must not take a day's leave without special permission from Rector and students; and if he left the city he had to pay a deposit which was forfeited if he did not come back. If he married, his honeymoon was limited to one day.

The law school at Bologna was largely independent of the other faculties that had grown up beside it to form the University. The law students were generally rather older than the others, many of them having already taken first degrees. They were grouped into "nations" roughly corresponding to the parts of Europe from which they hailed. Thus Copernicus was enrolled in the German "nation," one of the largest and the most privileged of them all. The original register containing his name has been preserved; it proves that he entered upon his studies at Bologna in the autumn of 1496.

From the old rules of Bologna University that have come down to us, it is possible to form some sort of idea of the everyday life of Copernicus as a student there. His day would be divided up by the church bells ringing for service. His first lecture might begin as early as seven o'clock in the morning, and last for two hours, students being expressly forbidden to show impatience by "drumming on their benches." The rest of the morning would usually be free; but after lunch there were classes for another three and a half hours. Besides general courses on the standard law books, there would be extra lectures on special books, or points of interest; also debates between teachers and students, or among the students themselves.

Canon law, then, was Copernicus's professed subject of study at Bologna. But his wide range of interests covered many other branches of knowledge as well—particularly mathematics and astronomy. In fact, the most important influence upon him at Bologna was that of the Professor of Astronomy there, one Domenico Maria da Novara (1454-1504). It is possible that Copernicus roomed in Novara's house. Normally, students as old as he was lived in lodgings at a controlled rent which was fixed each year by a committee made up of students and townsfolk. At any rate Domenico and Copernicus observed the heavens together, and they freely discussed plans for improving upon the old Ptolemaic system, and for simplifying it in various ways. Domenico da Novara was one of the leading spirits in the great revival of Greek studies which was just then sweeping through Italy and spreading into northern Europe, reaching even remote Cracow, as we have seen. The ideas behind this movement largely went back to Plato and even to Pythagoras. On the scientific side, they led men to try to picture the Universe by means of simple geometrical figures, or relations between numbers. And his friendship with one of the leaders of this movement must have encouraged Copernicus to go ahead with plans for reforming astronomy along these lines.

We know that Copernicus actually observed the heavens at Bologna from the fact that the earliest observation of his own which he mentions in his book of 1543 dates from this period of his career. It is an observation of the exact time at which the Moon was observed to pass in front of the bright star Aldebaran, hiding it from view (on March 9, 1497).

Copernicus was joined, late in 1498, by his brother Andrew, who had also come to Italy to study Canon Law. How far the brothers joined in the notoriously wild student-life of the city it is difficult to judge; but their expenses were considerable. Both were still largely dependent on their uncle. Towards the end of 1499, they ran out of cash. Happily, the Bishop's secretary, who was on a mission to Rome, was staying at Bologna. To him the brothers appealed, Andrew threatening to go to Rome and take the first job that offered if speedy help were not forthcoming. The secretary, indeed, was in no better case than they; but he put them in touch with the Ermland representative at the papal court, Bernhard Sculteti, who proved a friend indeed.

After three and a half years at Bologna, Copernicus was ready for a change. The Church was observing 1500 as a Jubilee Year. Priests and monks and devout Catholics flocked to Rome from all over Europe. Copernicus went there, too, with his brother in time for the Easter celebrations; very likely they acted in some sense as official representatives of the Ermland Chapter. They must have been present at the great ceremony on the Easter Sunday afternoon, when 200,000 people knelt in the open air to receive the Pope's blessing. It was during the visit of Copernicus to Rome that the Pope's son, Cesare Borgia, arranged the murder of his sister Lucrezia's husband, a deed for which he was never called to account. Ten years later, Martin Luther would come from Wittenberg to Rome, to return thence zealous for the reform of the Church.

Copernicus remained in the Eternal City for a whole year, giving informal lectures on mathematics and astronomy. From the period of his stay at Rome dates another of the observations which he used in his book. This time it was an eclipse of the Moon, on November 6, 1500.

While Copernicus had been studying at Bologna (it was probably in 1497) his uncle had put his name forward a second time for election as a canon of Frauenburg Cathedral. This time he was successful; though Copernicus did not actually enter upon his duties until years later, when he had finished his studies in Italy. His brother Andrew was similarly elected to a canonry in 1499. Early in 1501, the two young men paid a short visit home, and they were formally installed in the Cathedral Chapter on July 27 of that year. Immediately they applied for further leave of absence to resume their studies in Italy. This was granted after some discussion, the Chapter being favorably influenced by a proposal from Nicholas that he would undertake the study of medicine. Before the summer was over, the brothers had set off again on their travels. They crossed the Alps together and then parted, Andrew making for Rome, and Nicholas for Padua.

17. *Padua and Ferrara*

Throughout the Middle Ages, students were allowed, and even encouraged, to move from one university to another. Wherever they went, they could start their studies at the stage to which they had previously carried them. So when Copernicus returned to Italy, he sought out the University of Padua.

Here, too, there was a famous school of law. It had been founded in the thirteenth century by teachers and scholars who had migrated from Bologna. We are reminded in Shakespeare's *Merchant of Venice* of the fame of Padua as a center of legal studies. When Portia enters the law court disguised as a lawyer, she pretends to have come from Bellario, "learned doctor" of Padua whom the Duke has summoned to give his counsel on the suit of Shylock. Around the original law school, there had grown up a complete University. It was noted for its medical studies; and this may have been another reason why Copernicus was attracted to Padua, since medicine was now to be one of his principal subjects. Or perhaps both brothers felt it was time to break away from the distractions of Bologna. Copernicus would miss the friendship of Domenico da Novara; but his place may have been filled in some measure by the Paduan professor Girolamo Fracastoro, philosopher, physician and astronomer. He was a reformer in all these departments, and he may have helped Copernicus to break away from the Ptolemaic tradition.

Copernicus, then, completed his law studies at Padua; but for some reason he did not graduate there. He chose instead the University of Ferrara, where, in 1503, he was awarded the diploma of Doctor of Canon Law in the picturesque ceremony customary on such occasions.

A candidate for the doctorate had first to swear that he had been through the proper course of study. He was usually presented to the University authorities, or "promoted" as it was called, by a Doctor who would normally be his own teacher. The candidate appeared before the assembled Doctors who assigned him two passages in Canon Law. He retired to study them with the assistance of his "promotor"; and later in the day he appeared again before the Doctors to give an explanation of the two passages, after which the Doctors were free to question him on them. A vote was then taken by ballot as to whether the candidate should pass. If the vote was favorable, he became a *licentiate*, and proceeded (if he could afford the expense) to the public examination which made him a full Doctor. The private examination was the real test; if he passed that, the rest was a formality.

When the great day of the public examination arrived, the candidate proceeded to the Cathedral accompanied by his "promotor" and his fellow students. He there made an oration, and delivered a lecture on a point of law, defending his views against student opponents. He was then seated in the master's chair, and was presented with a cap, a book and a gold ring, together with a certificate or *diploma*. (Copernicus's diploma came to light not many years ago.) His "promotor" gave him the kiss of peace; and the proceedings were followed by a general jollification at the new Doctor's expense. The charges represented by this latter item were often so heavy that a candidate might prefer to take the final ceremony at some University other than his own, thus giving his convivial friends the slip. That may have been the reason why Copernicus graduated at Ferrara. After taking his degree, he lingered on for some months in the city. Magnificent buildings were going up there under the patronage of the Duke, Ercole d'Este. The Duchess was none other than the Pope's daughter, Lucrezia Borgia, who, after many adventures, had settled down at the early age of twenty-two to a life of good works.

Back again at Padua with his law classes finished, Copernicus was able to start work in earnest on the study of medicine. It was not that he intended to become a practitioner, and, indeed, he does not appear to have taken any medical degree. But in those days it was thought quite fitting that a Churchman should know something about the art of healing, so as to be able to minister to the physical as well as the spiritual needs of his flock. Otherwise, there was no skilled medical aid for the sick poor, only quacks and charlatans. In fact one of the conditions under which Copernicus enjoyed these further years of freedom and financial support was that he should acquire this knowledge. He was not expected to go in for surgery, which, in those days, before the invention of anesthetics, was a pretty ghastly business, calling for a greater degree of toughness than a mild Churchman could be expected to possess.

Unlike what we find in a modern medical school, medicine was then largely taught out of books. There were rules supposed to have been written by the Greek physician Hippocrates in the fifth century B.C.; and there were the writings of the Roman doctor Galen of the second century A.D., and of Avicenna, the Arab chemist and medico of the eleventh century, with other lesser authorities. But practical anatomy was still generally regarded with the same mixture of distrust and repugnance with which psychical research, or spiritism, is viewed today. Dissection of the human body had long been banned altogether. Now it was merely employed to illustrate what was in the books. At Padua, as elsewhere, it was customary to utilize the corpses of executed criminals for this purpose. One teacher read out of the book, a second explained the passage, and then, assisted by two

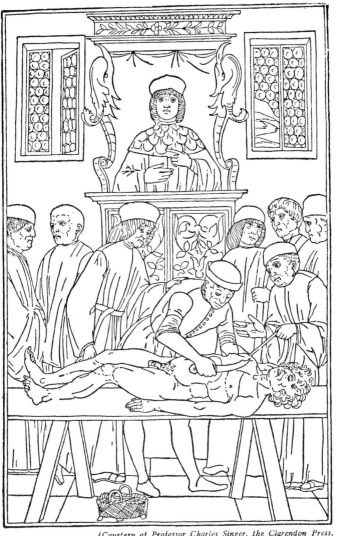

(Courtesy of Professor Charles Singer, the Clarendon Press, Oxford, and the British Museum)

FIG. 8. AN ANATOMICAL LECTURE AT PADUA
IN THE FIFTEENTH CENTURY

senior students, demonstrated it on the body (Fig. 8). At some Italian universities the teachers and students taking part in a dissection were fortified with a special issue of food, wine and spices. There was very little training in the treatment of actual patients, such as a modern medical student receives in walking the wards of a hospital.

The gulf between astronomy and medicine did not appear so great in those days as it does to us. For the sciences were not so specialized and self-contained as they are today. Moreover, in the Middle Ages there was supposed to be a mystic correspondence between the several parts of the Universe, and the organs of the human body. It was even supposed that the body was a small-scale model of the vast Universe. Old manuscripts that have come down to us have diagrams showing each part of the body connected with a corresponding Sign of the Zodiac —the head with the Ram, the feet with the Fishes, and so on. In a book published the year after Copernicus died, we are told that the brain corresponds to the sphere of stars, the eyes to the Sun and Moon and the ears, the nostrils and the mouth, to the other five planets. This, of course, was all part of the prevailing belief in astrology. Another link beween the two sciences was the elaborate set of rules as to the times of night, or of the month, in which medicinal herbs were to be gathered.

In a later Chapter we shall say something about the practice of Copernicus as a physician. But there was one other important event which marked this stage in the career of the astronomer. Some time during his stay in Italy, and probably during the Padua period, he learned Greek. This meant that he was now able to read the original works of Greek writers which had not yet been translated into Latin. And some of these writers helped to inspire and confirm the new ideas which were gradually crystallizing in his mind.

But now his leave of absence was running out; a war was creeping up Italy towards Padua, and it was time to be moving. At some uncertain date, but not later than the beginning of 1506, Copernicus was back in Ermland, his years of foreign travel and study over forever.

18. Heilsberg Castle

Copernicus returned from Italy a man in his early thirties, having had an education such as few enjoyed in those days. He had received the training in theology and philosophy which was required of a clergyman, and he had carried his studies in Church Law to the level of the doctorate. In the meantime, he had gained a fluent mastery of Latin, written and spoken, and a respectable knowledge of Greek. He had read many of the famous classics of Greece and Rome, which now abound in school editions, but were then the prize of the manuscript-hunter. Besides these literary classics, the ancients have left us great mathematical and astronomical books which few people read nowadays. Copernicus had studied deeply in these works, and they taught him all that was then known about these subjects, and served as the foundations upon which his own creative work was built. Lastly he had made himself acquainted with what passed for medical knowledge in his day.

After so many years of preparation, Copernicus might now at last have expected to settle down to his duties as a canon of Frauenburg Cathedral. But Bishop Waczenrode was ageing, and he wanted his nephew near at hand to keep him company, to assist him in governing his little domain, and to act as his medical attendant in case of illness. Perhaps, also, the Bishop hoped to save his brilliant nephew from consuming his time and energy in the petty duties which the Chapter would have laid upon its latest recruit. He may even have intended to train the young man as his eventual successor in the Bishopric. However, it was as private physician to the Bishop that Copernicus now obtained further leave of absence—for over six years, as it proved.

So Copernicus took up his abode at Heilsberg Castle, the official residence of the Bishops of Ermland. The Castle stood on the little river Alle, about 40 miles southeast of Frauenburg, in a picturesque, well-watered countryside. In those stormy times it had often suffered from fire and foe. As recently as when Copernicus was at Bologna it had been partly burned down. But Lucas Waczenrode had rebuilt it, grimly fortified without, but richly furnished within; and there he lived in almost royal state.

Ermland was the largest of the Prussian dioceses. The Order of Teutonic Knights owned two-thirds of the land to the Bishop's one-third. But, by his shrewdness and force of character, Lucas had kept his little state largely independent, while the other three dioceses had been swallowed up by the Order. There was a rule that mem-

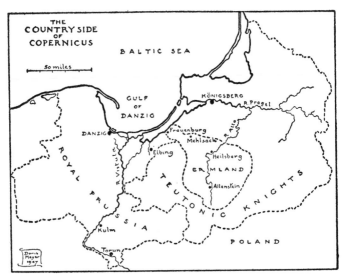

FIG. 9. THE COUNTRYSIDE OF COPERNICUS

bers of the Order were not to become canons of Frauenburg; otherwise the Chapter would soon have been swamped with Teutonic Knights.

For six years, then, Copernicus took part in the intricate business of governing the little state of Ermland, and of managing its relations with the neighboring states. On the one side were the Teutonic Knights, anxious to maintain and extend the power that still remained to them. On the other side was the Kingdom of Poland, equally anxious to grab all the territory that had belonged to the Order. Squeezed in between these quarrelsome neighbors, little Ermland had a hard task to survive at all. Lucas Waczenrode had had trouble with the Polish King over his consecration as Bishop; but that had been smoothed over, and now his influence was thrown to the Polish side. He accordingly incurred the hatred of the Order, which did everything possible to undermine his authority, and, as the story went, prayed daily for his death. On his part, Bishop Lucas had a plan for packing off the Teutonic Knights to fight the Turks; but it came to nothing.

After the Order had been forced to give up West Prussia, there was a conflict between the Prussian population of that area, who wanted to be completely independent, and the King of Poland, who wanted to incorporate them in his Kingdom. The Bishop had the delicate task of mediating between these two contending parties. It fell to Copernicus to represent Ermland at local Polish or Prussian parliaments, or *diets*, and to accompany, or act for, his uncle on diplomatic missions to Cracow or to Torun. It was also part of his duties to draw up memorials or addresses to the King of Poland, and to take the minutes of the proceedings in the Chapter. Copernicus also shared with his uncle in a scheme, which came to nothing, for founding a Humanist University at Elbing, in Prussia, for students who could not afford to study in Italy.

But at Heilsberg Copernicus found time for pursuits of a very different kind. For it was there that he began to set in order his scheme for refashioning astronomy on entirely new lines. We have seen how, as a student at Cracow, he had realized that the apparently complicated movements of the planets could be made much simpler by supposing that the Sun, and not the Earth, was the fixed central body about which the planets revolved roughly in circles. Later, in Italy, he had been strongly influenced by men who thought that the world of nature was really quite simple, and that, if it appeared complicated to us, that was because we did not understand it properly. Hence the simplest way of representing how the planets move must be the truest. Now, at Heilsberg, he began to set down his theory in writing. About 1512, or a little earlier, he wrote a short general account of his new system. Its Latin title means the "Little Commentary." Copernicus gave written copies of this short tract to a few trusted friends; but he did not publish it in book form. In fact it was not until towards the close of the last century that it was first printed—from two manuscript copies that have fortunately survived. It represents a definite stage in the gradual development of his ideas.

However, the "Little Commentary" was only a sort of first fruit, the token of a greater harvest to come. From now on we shall catch occasional glimpses of a great work to which Copernicus set his hand about this time. This was nothing less than building up his new scheme of the world into a complete system, numerically precise in every detail, and setting it forth in a book which was to become one of the classics of science. This task was to run like an increasing purpose through the duties and formalities of thirty years until at the last the new world-system would come to birth like a Phoenix from the mortal ashes of its creator.

For about six years Copernicus lived thus at Heilsberg Castle. Then an event occurred which brought to a close this phase of the astronomer's career. Lucas Waczenrode and his nephew had come to Cracow to be present at the wedding of King Sigismund, and the coronation of his young Queen. On the way back to Heilsberg, the Bishop was overtaken by sudden illness. He was only in the middle sixties, but he was already worn out by the strain of the anxious and responsible position he had now held for over twenty years. It was wintry weather, and the sick man was taken to the nearest place of shelter. It happened to be Torun, his birthplace; and there he died on March 29, 1512. His body was borne to Frauenburg and buried in the Cathedral.

It was said of Lucas Waczenrode that he was never known to laugh. And there is much to repel us in this hard-headed prelate, so zealous and cunning in his fight for the worldly power and prestige of his Church, so lacking in the tenderer Christian graces. Even Copernicus himself cannot always have found life easy with the self-willed old Prince of the Church. But he was rightly grateful to his uncle, as also we may be, for giving him, as the occasion of those times best served, the one kind of life in which his genius could flower and bear its fruit. Today Copernicus might have become a research fellow of a university or institute. In the sixteenth century, he became a canon of Frauenburg.

However, the death of Lucas Waczenrode marked the end of Copernicus's long residence at Heilsberg; and he now prepared to take up his abode at the Cathedral. By June 5, 1512, he was at Frauenburg, for he observed there, on that date, an opposition of the planet Mars.

19. Copernicus the Humanist

One of the ways in which Copernicus tried to show his gratitude to his uncle, the protector of his orphan boyhood, was by dedicating a book to him. It was the only book he ever brought out on his own account, and it had nothing to do with astronomy. It was a Latin translation he had made of the Greek verses of an old poet called Theophylactus Simocatta, who lived in the Byzantine, or Eastern Roman, Empire in the seventh century A.D. Theophylactus was not a great poet; and it is not easy to say why Copernicus chose his book for translation. Perhaps he had worked through it as a textbook while he was learning Greek, or perhaps he chose it because it was short and easy.

The book is a collection of eighty-five brief poems called *Epistles,* or letters, supposed to have passed between various characters in a Greek story. They are of three kinds—*moral,* offering advice about how people ought to live; *pastoral,* giving little pictures of shepherd life, and *amorous,* consisting of love-poems. They are arranged to follow one another in a regular rotation of subjects.

Copernicus translated the Greek verses into Latin

prose. His knowledge of Greek was imperfect, as was to be expected from the great difficulties which students of the language had to overcome in those days. Greek grammars and dictionaries were only just beginning to appear. Copernicus's copy of Chrestonius's Greek-Latin dictionary is preserved at Upsala, in Sweden, with notes in his handwriting, and his signature in Greek. He dedicated the book to his uncle, as an act of gratitude for all the benefits he had received from him. And in 1509, when the two men were in Cracow together on a political errand, Copernicus had the little book published at a printing press in the city. It was an event of some importance as one of the earliest attempts to introduce Greek literature into that bleak corner of Europe. Copernicus thereby proclaimed himself as on the side of the "humanists" in the struggle then going on over the question whether Greek literature ought to be revived or not. The "scholastics" feared that it would lead to a revival of ancient heathenism; and they were not far wrong so far as Italy was concerned.

However, the translation of Theophylactus attracted very little attention, and it was soon forgotten. It was not until the eighteenth century that copies of the book began to be discovered and treasured by collectors as the work of the great astronomer. One reason for this neglect was doubtless that the name of Copernicus did not appear on the title page but only at the end of the Dedication.

In those days it was the custom for an author to ask his friends for poems to put in the front of his book. Copernicus had a friend whose name was Rabe (meaning a *raven*), but who had changed it to Corvinus, from a Latin word meaning the same thing. He had been a wandering scholar who had taught for a time at Cracow, where Copernicus had attended his lectures. Now he had settled down as town clerk of Breslau, the capital of Silesia. Corvinus wrote some introductory verses for Copernicus's new book, in which he compared the astronomer's loyalty to his uncle with that of Achates to his friend Aeneas in Vergil's poem. He then speaks of Copernicus as "exploring the rapid course of the Moon, the various motions of

that star, the Earth's brother (the Moon), and of the
whole sky and the planets, the wonderful creation of the
Father of all. Starting from wonderful principles, he
knows how to explore the hidden causes of things." So
here we have, perhaps, the earliest suggestion of the lines
along which the mind of Copernicus was already working.

20. Frauenburg and Allenstein

The little town of Frauenburg stands on the coast of what
was once East Prussia, overlooking a lagoon of fresh water
which opens into the Baltic. The houses cluster round a
stately fourteenth-century cathedral built on a low hill.
From this rising ground one glimpses to seaward the
sandy spit which divides the lagoon from the darker
waters of the Baltic. To landward the eye rests on a rich,
well-watered countryside.

The Teutonic Knights had fortified the Cathedral hill,
a very necessary precaution in lands newly won from
heathendom. In the course of the wars between the Poles
and the Teutonic Order, the little town had changed
hands several times, for it was of some importance in war.
Finally a massive wall was built round the Cathedral for
its better protection. This wall was strengthened by turrets which, in peacetime, afforded extra accommodation
for the members of the Chapter, whose regular apart-
ments were within the wall. The younger canons were
usually allotted the least comfortable quarters; they
moved into better ones as the older members of the
Chapter died off. Copernicus, however, seems to have
lived all through the thirty years he spent at Frauenburg in the same rooms. They were in one of the turrets,
which served him as an observatory.

In the time of Copernicus, the canons of Frauenburg
numbered about sixteen, including five "prelates" who
were charged with special duties. They were all under
obligation to "keep residence," that is, to live near the

Cathedral, although in practice, as we have seen, leave of absence could be obtained for all sorts of reasons. It was their duty to conduct the services, to hold morning and evening devotions, and to recite masses. They were also to advise and assist the Bishop in carrying on the spiritual and temporal business of the diocese. They were allowed, and were even required, to live in a certain style, with at least two servants and three horses apiece. Their income was derived from the ownership of land. The Bishop owned one-third of Ermland, and one-third of this in turn was for the benefit of the canons. Each brother had also his own freehold in the neighborhood of the Cathedral. The spiritual tone of the Chapter was low. The canons generally shrank from taking priestly vows, except those who aspired to be prelates or bishops. In fact, in Copernicus's time, the Frauenburg Chapter could barely muster one priest for the service of the altar.

When Copernicus settled at Frauenburg, he would already be well known there from his frequent visits to the Cathedral. Some of the canons were his relations, the Chapter being largely drawn from the leading families of Danzig and Torun who intermarried extensively. His brother Andrew was already living there; but Copernicus did not enjoy his society for very long. Andrew had developed leprosy, or some such disfiguring disease; perhaps he had picked it up as a student in Italy. His brother tried his medical skill on him, but to no purpose. In 1508, the invalid had been granted a year's leave of absence to travel in search of a cure; but in vain. In 1512, the year Nicholas went to live at the Cathedral, Andrew was forced to withdraw from the common life of the Chapter for fear of infecting his brethren. He died abroad, probably not later than 1519.

During his first few years at Frauenburg, Copernicus does not appear to have been overworked. It is probably from this period that we must date the original draft of his great book, and the earliest of the several revisions that the manuscript underwent before its publication in 1543, some thirty years later. But, as the years passed, he was increasingly pressed by the responsibilities for which he was marked out by his intimate knowledge of the peculiar situation of the little state of Ermland.

Right at the start the question arose, Who should succeed Bishop Waczenrode? And the yet more delicate question, Who should have the choice of his successor? The King of Poland claimed the right to choose the Bishops of Ermland, and he was backed up by the Pope. In the end, the King agreed to the Chapter's choice of one of their own number, Fabian von Lossainen, as the new Bishop. But he insisted that, in future, the Chapter should submit a list of possible candidates to the King, who would then pick out four, any one of whom would be agreeable to him. Copernicus himself was included on the short list for one of the later vacancies of the bishopric; but it was only a formality. The Chapter feared that the King would some day appoint a Polish Bishop, as indeed came to pass, though not until after the astronomer's death.

Towards the close of 1516, Copernicus was appointed to manage two outlying estates belonging to the Chapter. He was made responsible both for the local government and for the administration of Church affairs at Allenstein and Mehlsack. His headquarters were at Allenstein Castle, about 50 miles inland from Frauenburg, on the same river Alle that flowed past Heilsberg. From there he used to ride out with his servants to make the circuit of his territory. An interesting relic of this stage of the astronomer's career has come down to us. It is the book in which he recorded all the transactions involving the transfer of property on his little domain.

Copernicus would also come to Frauenburg from time to time to see his friends, to attend meetings of the Chapter, and, if opportunity offered, to observe the heavens from his little turret observatory. As many of the observations of Copernicus belong to this period, and as they were soon to be interrupted by the outbreak of war, this may be a convenient point at which to attempt some account of his work as a practical astronomer.

21. *Copernicus the Observer*

When Copernicus settled at Frauenburg, he chose for his living-quarters one of the turrets on the defensive wall which enclosed the Cathedral. "Copernicus's Tower" is identified by an old tradition as the turret which stands at the northwest corner of the enclosure (Plate VII). He must have chosen the place as well suited for an observatory, commanding as it did an uninterrupted view in all directions except to the east. The Tower has a basement and three floors. There used to be a door opening out of it onto the top of the wall, where there was a platform from which open-air observations could conveniently be made. Ever since the seventeenth century, "Copernicus's Tower" has been carefully preserved as a hallowed spot. It contains a portrait of the astronomer; and it was recently being used as the Cathedral Library.

Frauenburg was not a very good place for an observatory. It lies so far north of the equator that the planets always appear far down towards the southern horizon. They have to be viewed through a great thickness of air, filled with the vapors rising from the sea and from the well-watered plain in which the town stands. But it had the advantage of lying nearly on the same meridian as Cracow, a capital and university city to which, as a point of reference, observations could fittingly be referred.

The making of astronomical observations was an essential part of the work of Copernicus. For his theory was not just a vague suggestion such as an early Greek philosopher might have thrown out. It was meant to stand comparison with exact measurements of the positions of the heavenly bodies at definite times. Such measurements had been made by the Greeks and the Arabs. Many of their results had been handed down and were available to Copernicus. As a matter of fact, many of these old observations were worth very little. They had been made with crude instruments, and some of the results were bogus. And there was always the danger that, in copying them from one manuscript to another all down the ages, slips had occurred. Copernicus took them all on trust, at their face value. He gave himself a lot of unnecessary trouble, and he made his theories needlessly complicated, by trying to make them agree somehow with all the observations, good, bad and indifferent, that had come down to him.

Nevertheless, there was an idea abroad that the heavens

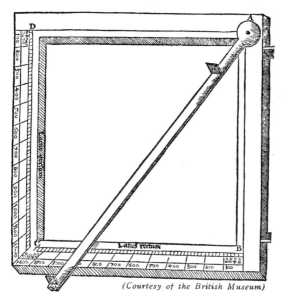

(Courtesy of the British Museum)

FIG. 10. AN ASTRONOMICAL INSTRUMENT OF
THE AGE OF COPERNICUS

The wooden quadrant of George Purbach (c. 1450). The movable arm, free to turn about a pin at the top right-hand corner, is directed by means of the two sights toward a star, the tangent of whose altitude is then read off the graduated sides of the quadrant.

had undergone slow changes with the lapse of the centuries. Hence Copernicus felt that he ought not to rely merely on these old observations, but must make a fresh set of measurements on which to base his theories. That was why he wanted an observatory.

It must not be supposed, however, that Copernicus was

a great observer, and that that was why we give him an honored place among astronomers. His instruments and methods were cruder, and his measurements less exact, than those of the best of the Greek astronomers more than a thousand years before his day. He had no exaggerated opinions about his own capacities in this matter. We know that it is possible, without the aid of a telescope, to measure an angle in the sky with an error of only about one or two minutes of arc. (The apparent breadth, or "angular diameter," of the Sun or Moon in the sky represents an angle of about thirty minutes of arc at the eye, as shown, with a great exaggeration of this angle, in Fig. 11.) But Copernicus told his friend Rheticus that if

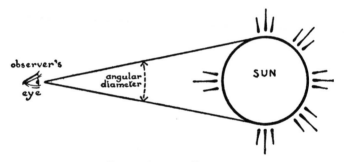

FIG. 11. ANGULAR DIAMETER

he could measure such angles correct to *ten* minutes he would be as happy as Pythagoras must have been when he discovered the proposition about the squares on the sides of the right-angled triangle.

In his book of 1543, Copernicus describes the several kinds of instruments employed by Ptolemy, and still occasionally used in his own time. But he does not make it quite clear whether he himself possessed all these types. However, he seems to have made at least two of these instruments, following the descriptions of them given by Ptolemy.

The first of them was intended for measuring the meridian altitude of the Sun. This is the angle which the direction of the Sun makes with the horizontal at the time when the Sun is due south, and therefore at its greatest elevation above the horizon. The instrument was a flat square slab of stone or metal AB (Fig. 12) set upright so as to lie north and south. With the corner A as center, and radius equal to the side of the square, a quarter of a circle was drawn and divided up in angular measure from 0° to 90°, like a protractor. From A a short rod AC stuck out at right angles to the slab. As the Sun S passed due south it cast the shadow of the rod on to the graduated circle. The reading on the circle gave the angle SAH,

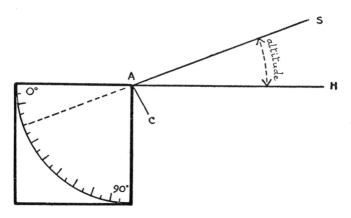

FIG. 12. MEASUREMENT OF THE SUN'S ALTITUDE

which was the Sun's meridian altitude above the horizon. Many important results in astronomy depend upon the measurement of this angle at various times of the year.

The second instrument served for measuring the altitude of bodies anywhere in the sky. It consisted of three wooden rulers, two of them about four yards long, the third rather longer, and jointed together (Fig. 13). The first of these, AB, was fixed upright, so that it pointed up to the *zenith* Z, that point in the sky which is vertically over the observer. The second ruler AC, was jointed to the ruler AB at A, so that it could turn about A through any angle BAC. It had a pin at C. The third ruler, BD, was jointed to AB at B; it was graduated in units of length, and each graduation had a hole into which the pin at C could be inserted so as to fix the ruler AC in position. There were two sights on AC by means of which that ruler could be pointed accurately at the particular star S whose altitude was required. The ruler BD was then pinned at C and it served as a crosspiece to hold AC and keep it pointing at the star. From the number of

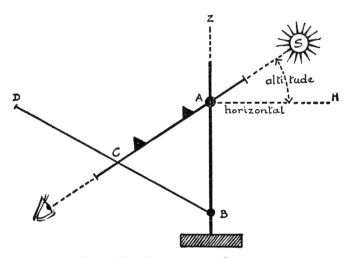

FIG. 13. THE TRIQUETRUM OF COPERNICUS

graduations between B and C, the angle BAC could be calculated. This was equal to the opposite angle ZAS; and when this was subtracted from 90°, the required altitude SAH was obtained as before.

This *triquetrum*, as it was called, was kept at Frauenburg for about forty years after the death of Copernicus. Then, in 1584, Tycho Brahe, the great Danish astronomer whom we shall meet in a later chapter, sent one of his assistants to measure the latitude of Frauenburg. One of the canons gave this assistant the instrument to take back as a present to Tycho; he valued it highly and took it with him when he moved from Denmark to Prague. After Tycho's death, the Emperor Rudolph II had it. But later Prague was sacked in one of the religious wars of the period, and Copernicus's *triquetrum* disappeared forever.

In his book of 1543, Copernicus makes use of 27 of his own observations. Of these, one was made at Bologna and one at Rome (as we have seen); probably all the rest were made from the turret observatory of Frauenburg. These observations were of such things as eclipses of the Moon, the elevations of the Moon above the horizon, the passages of the Moon in front of bright stars, or the positions of planets in relation to the background of stars. Copernicus required exact measurements of the times and circumstances of such events as the raw material out of which to build an accurate theory of the planetary motions. Besides the twenty-seven observations mentioned in his book, Copernicus also made others, the results of which he jotted down on the margins and flyleaves of the volumes which made up his little private library, where they can still be seen.

22. The Reform of the Calendar

We have already pointed out the valuable service which astronomy renders to mankind by furnishing us with an accurate calendar. This, however, was achieved only about 350 years ago after centuries of confusion. A solution of the problem was still being sought when Copernicus was living; and it was one of the matters to which he turned his attention.

The calendar which was in use in the early sixteenth century was the one which Julius Caesar introduced in the first century B.C. He gave up the attempt to make the length of the year depend upon the Moon as well as upon the Sun; and that saved a lot of trouble. He also made the normal length of the year to be 365 days, an extra day being added to February every fourth year (our *leap-year*) so as to give the year an average length of 365¼ days. This is very nearly exact, but not quite. For the length of the year is about 11 minutes less than this; and the difference adds up from year to year to produce an error of one whole day in 128 years. Moreover, for the purpose of fixing the date of Easter (which partly depends on the Moon) the Church assumed a relation between the length of the year and that of the month which was also not quite correct.

In the time of Copernicus, the accumulated results of these errors were becoming serious, and the Church authorities were alarmed. In 1514 they were meeting in a general Council; and the Pope, Leo X, took the opportunity to send out an SOS to the Christian princes and universities, as well as to private individuals, asking for their help in putting the calendar to rights.

One of the people invited to Rome for this purpose in 1514 was Copernicus, who was evidently becoming famous for his skill in such matters. The invitation was sent to him, with a personal plea, by Bernhard Sculteti, once his deliverer in student days (Chap. 16), now his brother-canon and domestic Chaplain to the Pope. But Copernicus rejected the invitation on the ground that the problem of the calendar could not be solved until the laws of motion of the Sun and Moon had been thoroughly cleared up. This had not yet been done, he explained, but he was addressing himself to the task.

Thirty years later he justified the publication of his book by the hope that it would help towards the reform of the calendar; and he dedicated it to the reigning Pope, Paul III. By then the Church had dropped the matter for the time being. But the determinations of Copernicus provided the foundation for the reform that was eventually carried out by Pope Gregory XIII in 1582. Gregory's remedy for the error in Julius Caesar's calendar involved omitting ten days from the year 1582, and thereafter omitting the extra day from leap year three times every four centuries. This adjustment was not accepted in England until 1752.

23. War

While Copernicus was administering the Allenstein estates, another war was brewing between Poland and the Teutonic Knights. As he rode the rounds of his little domain, or slipped off to Frauenburg to observe an eclipse, armed bands of Germans, recruited by the Order, were amusing themselves by murdering and plundering the defenseless peasants.

Open war broke out late in 1519, just after Copernicus had completed his third year at Allenstein and was back at the Cathedral. The Polish King had summoned the head, or "Grand Master," of the Order to meet him at Torun, and, when he failed to appear, broke off relations and marched into his territory. There were no big battles, each side preferring to kill noncombatants. Eventually the Grand Master suggested that the Bishop, Fabian von Lossainen, should mediate between the two sides, and invited him to Braunsberg, the Ermland town which the Knights had just captured.

The Bishop had always been a weak character, and now he was stricken with a mortal disease. Instead of going himself, he sent two of the canons. One of these was very likely Copernicus, for a letter of the Grand Master's has come down from this period promising to "the worthy, most learned Sir Nicholas Koppernigk, our free, safe, and Christian conduct into our Order's land" with his servants and horses. But when the Grand Master demanded an oath of allegiance from Bishop and Chapter as part of the peace terms, the two emissaries explained that they would have to make further inquiries about that. They left Braunsberg, and the peace talks came to nothing.

Copernicus spent the year from November 1519 to November 1520 at Frauenburg. He even resumed his observations there, safe behind the Cathedral wall while the little town went up in smoke. Most of the other canons had taken fright at the war, and he had the place almost to himself.

When the war turned against the Teutonic Knights, and the Poles had come to terms with Moscow, the Grand

Master asked for an armistice and went to Torun. But he broke off the negotiations when he heard that Germany was sending him troops. Marching against Heilsberg, he besieged the town and bombarded the Castle.

Meanwhile, Copernicus had resumed his duties at Allenstein Castle, under difficult circumstances. The countryside was devastated, the peasants slain or scattered abroad. Most of the canons had evacuated themselves to places of safety. Some of them had fled as far as Danzig, where they lived at ease and haggled over the sharing of their income. Only two canons besides Copernicus seem to have remained in Ermland. One of these, Henry Snellenberg, was with him at Allenstein Castle, where, for a time, they were besieged, though not very closely, by the forces of the Order. The other canon, old John Sculteti (not to be confused with Bernhard), sent to Copernicus from Elbing some of the crude little guns of the period, promised him powder and shot if he needed them, and urged him, somewhat unnecessarily, not to surrender the Castle.

Allenstein held out; both sides wearied of the war, and it ended in 1521 in an armistice. This was to last for four years in the first instance; but it afforded a chance for a permanent settlement. The whole region became more settled and peaceful in 1525, when the Grand Master of the Teutonic Knights agreed to become a Duke. He took this step on Luther's advice, and soon after turned Protestant. He was allowed to retain his little state of East Prussia in return for doing homage to the Polish King.

Meanwhile Copernicus was trying to repair the damage which the war had done to Ermland. Being neutral, the little state had been a battleground for both sides. The tenants had nearly all fled from the Allenstein estates, leaving them to go to rack and ruin. They had to be collected together again and resettled on the land, or new tenants found. Until the Chapter reassembled, Copernicus was "Commissioner for Ermland," and found himself in charge of the whole concern. He was also entrusted with the task of drawing up a detailed account of all the damage the Order had done to Ermland, and of presenting it at the peace-conference. In this task, Copernicus was assisted by a brother-canon, Tiedemann Giese, one of the best friends he ever had, of whom we shall soon hear more.

However, Copernicus also took advantage of the lull following the war to give another revision to the bulky manuscript in which his new system of the world was gradually taking shape. The rediscovery of the manuscript during the last hundred years has enabled students to form some idea of the slow, groping process by which Copernicus built up his book, adding a passage here, crossing out something there, and sticking on a bit somewhere else. By now the work was nearing completion, and rumors of its import were beginning to get about. Yet Copernicus was still not satisfied with it; and the expectations of the learned world were to remain unfulfilled for another twenty years.

24. The Diseases of Money

The typical scientist is sometimes represented as a bald-headed old gentleman with spectacles and a butterfly net. Copernicus was not that kind of scientist. He was a capable man of affairs. He had a deep understanding of the intricate machinery by which a country is run, both in its inner life, and in its outward relations to other countries.

A most important part of that machinery is money. It is like the electric current which keeps all the wheels turning in a vast factory, or like the lifeblood coursing through the arteries to vitalize all parts of the body. Money has its own mysterious laws and properties which we cannot alter but can learn to understand and use just as we do the laws of the electric current, or of the healthy working of our bodies. There is a science of money; and one of the men who founded it was Copernicus. Money has its diseases. And Copernicus, who had never taken a college course in economics, understood and prescribed for those diseases much more efficiently than for the human maladies which he had spent so long learning how to treat at Padua.

Money enables us to exchange the things that we make, or the services we render, for the goods that we really want. Without it we should have to dig the butcher's garden when we wanted a leg of mutton, or make the baker a pair of shoes in return for his bread. Money also makes it possible for us to compare the values of very different things, or "commodities," with one another, or the values of the same thing at different times, by expressing them all in the same units, dollars, or pounds or francs. But this process can be reversed; and instead of saying that so many bushels of wheat are worth so many dollars, we can say that so many dollars are worth so many bushels of wheat. For money, too, is a "commodity"; its value can rise or fall, with very important consequences to the individual and to the nation. When the value of money falls, everything in the shops gets dearer and dearer. People then demand higher wages; but that may only make things dearer still.

Nations have had to face this danger of what is called "inflation" following the recent war. The same thing happened in sixteenth-century Prussia, following the long series of wars between Poland and the Order. It forced Copernicus to think about money; and in the dark evenings at Allenstein Castle, he wrote a little book about it.

The trouble in Prussia was that the coinage had become very debased, the various authorities seeking a temporary financial advantage by putting less and less gold and silver into the coins that they issued. This drove up prices and made foreign trade almost impossible, as foreigners would not take the worthless coins in exchange for their goods. Common action was called for to put things right; but each state—the Poles, the Prussians, the Order, and even each big city, such as Torun and Danzig—clung jealously to its own right to coin its own money. Attempts to get agreement had always broken down.

Copernicus said the ruler or the government that tried to gain an advantage by debasing the coinage was like a farmer who should try to make a profit by sowing bad seed because it was cheaper than good seed. He was all in favor of a *monetary union* of all the states concerned: they should all agree to have the same coins. There was to be only one authority issuing coins; the amount of money in circulation should be limited; each coin was to contain a guaranteed weight of precious metal, and the debased old coins were to be called in. Copernicus was aware of the practice of exchanging bad coins for good and melting down the latter or sending them abroad. This tendency of bad money to drive good money out of circulation came to be called "Gresham's Law" after a later discoverer, Sir Thomas Gresham.

Copernicus seems to have drawn up some notes on these lines while he was at Allenstein in 1519. He made them the basis of a report on the matter, written in German, which he presented to the Prussian Diet held in 1522 at Graudenz, whither he had traveled with his friend Tiedemann Giese to represent the Chapter. He later drew up a revised and enlarged version of his little treatise, this time in Latin, and setting forth a general theory of money, for presentation to the Diet of 1528. The agitation for reform dragged on. The statesmen listened to the astronomer's advice. But the opposition of those who were profiting by the existing state of affairs was too strong, and nothing came of it.

About this time, also, Copernicus devised a scheme for a bread tax. It was intended to serve the purpose of keeping the cost of living steady, and of avoiding the frequent alterations in the price of bread which were responsible for a lot of hunger and poverty. He also made several tours of inspection of the Chapter's estates at Allenstein and Mehlsack, to see how they were settling down after the war.

25. Later Years

The scheme for reforming the coinage, and special assignments from the Chapter, continued to keep Copernicus in touch with public affairs after the war was over.

But as he grew older, his official duties gradually passed into the hands of younger men. He was thus free to concentrate to an increasing extent on his scientific work. He went on observing the heavens, looking out for interesting celestial events, eclipses, conjunctions of planets with stars, and so forth. And he continued to busy himself with the revision of his manuscript.

There was an early tradition that Copernicus observed the great comet of 1533. He was said to have become involved with other astronomers over the question of why this object moved round the sky the opposite way to the planets. We now have reason to believe that he actually wrote a little tract on this comet.

However, it is to be feared that the later years of the life of Copernicus must have been for him a time of increasing loneliness. One by one the friends of his youth were carried off by death, and their places in the Chapter were taken by men with whom he had very little in common. This growing isolation was partly the result of changing conditions in the Church.

While the war had been occupying everyone's attention in Prussia, the Reformation had been making great strides in Germany. The very year, 1521, in which the armistice was signed was that in which Luther made his defiant stand at the Diet of Worms. The Protestant doctrines quickly spread to Prussia. They secured a strong foothold in trading cities such as Danzig. The Duke of Prussia, as we have seen, became a Protestant. And the reformed religion spread even in Poland; though later, under the preaching of the Jesuits, Poland became one of the most Catholic countries in Europe. Bishop Fabian von Lossainen, the successor of Lucas Waczenrode, was not unsympathetic to the reforming movements. Copernicus himself does not appear to have left on record any direct expression of his own views on the matter. But he did much to inspire a book written by his friend Tiedemann Giese, and published in 1525, which probably expresses his attitude.

Tiedemann Giese had been a member of the Chapter for about twenty years. At the time when Copernicus went to live at Frauenburg, it was Giese's turn to be away administering the Allenstein estates; but after his return to the Cathedral he became one of the closest friends the astronomer ever had. Born in 1480 of a Danzig family, Giese had studied at German and Italian universities, and he was to rise high in the service of the Church, first as Bishop of Kulm in Prussia, and later as Bishop of Ermland. In 1549 he joined his friend, already laid to rest in Frauenburg Cathedral.

Giese was a convinced Catholic, but broad-minded and tolerant, and he kept on good terms with Melanchthon, the Protestant leader. However, the book to which we are now referring was a reply to one written by a Lutheran Bishop. In it, Giese states that Copernicus supports the views he is expressing, and has urged him to publish the book. Both men were anxious to preserve the unity of the Church, and to avoid the split between "Catholics" and "Protestants" which has continued to our own day to weaken seriously the witness of the Church in the world. They called for a more Christian spirit in preserving this unity. Catholics should plead or argue with Protestants instead of burning them alive as the King of Poland was doing in Danzig. "The wild animals behave more gently to their kind than do Christians to theirs," wrote Giese. Copernicus was aware of the evils in the Church life of his time. He had had ample opportunities to study them at close quarters. But he was anxious to preserve what was of value in the old form of the faith—not to throw away the baby with the bath-water—and not to unsettle the beliefs of simple folk.

In 1523, however, the moderate Bishop Fabian died. There was the usual rumpus and period of confusion before a successor could be appointed, and for six months Copernicus acted as Administrator-General of Ermland. When the new Bishop was announced, he proved to be a member of the Chapter, Mauritius Ferber. He figures somewhat prominently in Reformation history as a bitter foe of the Protestants. Most of the canons took the same line as their Bishop. Copernicus was more tolerant; and that was one of the reasons why he was increasingly left to himself in his declining years.

Another reason for this isolation was, of course, the

novelty of his opinions concerning the solar system. A generation or so earlier, these would not have been taken so seriously. But, following the revolt of the Reformers, the Catholic Church showed a growing distaste for any more departures from the old established beliefs, even concerning things which we should think had nothing to do with religion.

Since Catholics showed such a dislike of the teachings of Copernicus, we might have supposed that therefore their opponents, the Protestants, would have embraced the new theory eagerly. But actually the Protestants rejected the Copernican system even more decisively than the Catholics. Copernicus had been in his grave for many a long year before the Catholic Church made an official pronouncement that his teachings must be rejected. But even before the publication of his book, Copernicus had been severely criticized by Luther, who, in conversation, used to denounce "the new astronomer who wants to prove that the Earth goes round, and not the heavens, the Sun, and the Moon; just as if someone sitting in a moving wagon or ship were to suppose that he was at rest, and that the Earth and the trees were moving past him. But that is the way nowadays; whoever wants to be clever must needs produce something of his own, which is bound to be the best since *he* has produced it! The fool will turn the whole science of Astronomy upside down. But, as Holy Writ declares, it was the Sun and not the Earth which Joshua commanded to stand still." Luther's colleague, Philip Melanchthon, also argued against the Copernican theory in a little book on physics which he published some six years after the astronomer's death. Like Luther, he laid especial stress on texts in the Bible which seemed to suggest that the Earth was at rest.

These remarks of Luther and Melanchthon show how, even before the publication of his great book, the teachings of Copernicus had become known to the learned men of his day. But the rumor of his opinions had also reached the ears of uneducated and pig-headed people who merely poked fun at them. Opposition to his teachings at this low level took the form of a burlesque play which was produced about 1531 at Elbing in Prussia, not far from Frauenburg. Little is known about this play except that it was written for a carnival by a local schoolmaster. It seems to have held Copernicus up to ridicule by introducing a thinly disguised representation of the star-gazing priest into a procession of grotesque figures. But the whole episode may have had the wider purpose of mocking at the canons of the local Cathedral. And some have even thought that the Teutonic Knights were at the back of it.

Bishop Ferber died and was succeeded, in 1538, by the Bishop of Kulm, who called himself Dantiscus. He had coined the name from that of Danzig, where he had been born, the son of a brewer, in 1485. His place at Kulm was taken by Tiedemann Giese, who thus became a Bishop. Dantiscus was still Bishop of Ermland when Copernicus died. In his younger days he had led a wild life as soldier and student all over Europe and the Middle East. Now, as an elderly priest, he was trying to make up for it by being absurdly strict where other people were concerned. He continued the policy of ruthlessly repressing the Lutherans. He fell foul of Copernicus, partly because Copernicus showed friendship to the Protestant scholar Rheticus (of whom we shall speak in a later chapter), and partly for other, more personal reasons. Dantiscus also took measures against Alexander Sculteti, a canon and a friend of Copernicus, who had been elected to a prelacy in the Chapter which Dantiscus wanted for one of his own friends. The quarrel dragged on for years; in the end Sculteti was supported by the Pope; but Dantiscus expelled him as a heretic, and had him hounded out of Prussia.

However, the later years of Copernicus were not without their encouraging experiences. The "Little Commentary" (p. 74) had made his teachings generally known among scholars. And here and there the seed fell upon kindly and receptive soil. The little manuscript seems to have provided the material for a lecture on the new theory which John Widmanstad, the papal Secretary, gave in 1533, before Pope Clement VII and some of his cardinals in the gardens of the Vatican.

Again, in 1536, Cardinal Nicholas von Schönberg (whose secretary Widmanstad had then become) wrote to Copernicus from Rome begging him to publish the details of his system, or at least to communicate them to the writer. Whether Copernicus replied or not we do not know; and in any case von Schönberg died in the following year. But he prized the Cardinal's letter highly, and it was printed at the beginning of his book of 1543.

Copernicus was also greatly encouraged, in the last years of his life, by the visit of a disciple, Rheticus, who came to him from a very unlikely quarter—the Lutheran University of Wittenberg. This was in some sense the final and decisive event in the whole career of Copernicus.

However, what the people around him were chiefly aware of at this period was his growing fame as a physician. And this may be a convenient point at which to set down the few facts we know about this aspect of his many-sided activities.

26. Copernicus the Physician

It is hard to realize that, in his lifetime, Copernicus was more highly thought of, and perhaps better known, as a medical man than as an astronomer. And this in spite of, or perhaps because of, the fact that he blindly swallowed the opinions of his time on medical matters. Yet in those days both the theory and the practice of medicine were almost unbelievably crude and backward, and far more in need of reform than astronomy.

Nowadays we have a fair knowledge of how the human body works; and upon this knowledge we base our treatment of diseases. But in the sixteenth century, the most mistaken ideas still prevailed as to the structure and the working of the body. Even such an important fact as the circulation of the blood was not discovered until about a century later. In the meantime, the arteries (which carry the blood from the heart) were thought to serve

(Courtesy of the Mundus Publishing Association)

I. Nicholas Copernicus

II. CENTRAL EUROPE IN THE TIME OF COPERNICUS. From an Atlas of 1508

Weypel Aus

A. S Johannis Kirch
B. S. Jacobs Kirch
C. S. Marien Kirch u. Gymnasium
D. Dominicaner Kloster
E. S. Laurentz Kirch

G. Alt Staꝛtische Rathauss
H. Neu Staꝛtische Rathauss
I. Das Schlofs Barg
K. Das Dominicaner Thor
L. Euwicker Thor
M. Cuolmisches Thor

N. Holl Thor
O. Stobsr Thor
P. Schlofs vorr
Q. Die Jnfel

III. TORUN, THE BIRTHPLACE OF COPERNICUS

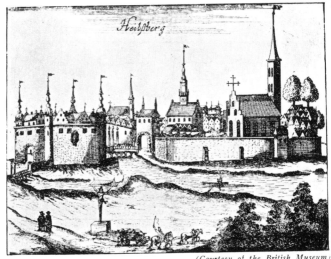

IV. HEILSBERG

V. FRAUENBURG

VI. ALLENSTEIN CASTLE

VII. COPERNICUS'S TOWER

merely for conveying air, or imaginary "vital spirits," about the body.

There was no idea of basing treatment even upon what was supposed to be known about the human frame. The usual type of treatment was to dose the patient with medicines prepared by mixing together a surprising assortment of materials. These included such things as goat's blood, stewed bats, the ground-up horns of animals, the lungs of vipers, spiders' webs, roots and barks, powdered jewels, and other even less appetizing ingredients! The body was scarcely regarded as a whole. Diseases were classified and remedies ordered according to the particular organ in which the symptoms chiefly appeared. And even this treatment was restricted by all sorts of astrological rules, because of the supposed connection between the Signs of the Zodiac and the several parts of the body or the seasons of human life (Chap. 17). For example, it was thought inadvisable to bleed a patient when the Sun was in the Sign of the Bull. Copernicus possessed a number of the standard medical books of his day, setting forth this sort of treatment. Many of these well-worn volumes have come down to us with his favorite prescriptions marked or copied in his handwriting in the margins.

As to his actual medical practice, we know very little. We are not told anything about his treatment of his uncle, his brother, and the other members of the Chapter in his younger days. However, the Cathedral records covering the later years of his life give brief notices of his attendance on the elderly Bishops who in turn occupied the see of Ermland.

Mauritius Ferber was often on the sick list with colic and gout from 1529 to 1537, when a stroke finished him off. Copernicus was constantly being summoned to Heilsberg. In treating such an important patient, he thought it wise to take other doctors into consultation. He called in the physician of the Duke of Prussia, and asked the advice of the Polish Royal Physician by letter. In those days, when traveling was difficult, it was not uncommon for doctors to treat their patients by mail. Several times we find Copernicus giving professional advice to his patients, or seeking the counsel of other medical men in difficult cases, in this manner.

He next attended the succeeding Bishop, Dantiscus, who had been taken ill while officiating at the marriage of the Crown Prince of Poland. A patient in whose recovery he must have felt a livelier interest was his old friend Tiedemann Giese, now Bishop of Kulm in Prussia. He had caught ague, a sort of malarial fever, while traveling in the marshy country of his diocese. Copernicus visited and treated him early in 1539. Later in the year, he made a second medical visit to Giese, which developed into a regular summer holiday. Accompanied by his young disciple Rheticus (of whom we shall speak in the next chapter), he went to Kulm in May and returned home to Frauenburg in September.

In the spring of 1541, two years before his death, Copernicus was hastily summoned by no less a person than Duke Albert of Prussia (the former Grand Master). He was to go to Königsberg to attend George von Kunheim, a trusted counsellor of the Duke's, who had fallen seriously ill, and for whom the Prussian doctors seemed unable to do anything. Copernicus went willingly enough, for he had met von Kunheim in the course of negotiations over the reform of the coinage. And he had come to feel that even Albert was not such a bad sort. The two had many intellectual interests in common. The Chapter readily gave Copernicus permission to go, as it wished to remain on good terms with the Duke, despite his Lutheran faith. The courtier's illness was prolonged, and the leave of absence had to be extended over Eastertide. At length, however, Copernicus was able to return, leaving von Kunheim well on the way to recovery; though he continued for a time to receive reports on the patient's condition, and to send him medical advice by letter.

There was an early tradition, which has caught the imagination of later writers, that Copernicus exercised his medical skill for the benefit of the local poor, who venerated him as a "second Aesculapius" (a Greek demigod of healing); but of this we have no further details.

27. *The Coming of Rheticus*

In the spring of 1539, a young German scholar arrived at Frauenburg and asked to see Copernicus. He was George Joachim, Professor of Mathematics at the recently founded University of Wittenberg. He had come, without a word of warning or introduction, to seek fuller information concerning the new astronomical system of which he had heard reports.

Joachim was then twenty-five years of age. He had been born in 1514 at Feldkirch in the Austrian Tyrol, a part of Europe which the Romans had called *Rhaetia*. He had therefore adopted the name of Rheticus, "the man from Rhaetia," by which he is generally known. Philip Melanchthon, the Protestant scholar, had picked him out as a clever young mathematical student at Wittenberg, and had made him a Professor there at the age of twenty-two. The other mathematical Professor was one Erasmus Reinhold, who also comes into our story.

Rheticus took a certain risk in coming from Wittenberg, a hotbed of Protestantism, into a Catholic diocese where Protestants were being ruthlessly persecuted. He was in danger, too, of getting into trouble with the heads of his own University, who strongly opposed the teaching of the Earth's motion. Copernicus, also, might have suffered for entertaining a heretic; perhaps it was one of the reasons for his friendlessness at the last. However, Copernicus cordially welcomed the young man, took him to see Tiedemann Giese at Kulm, and introduced him to other friends. Rheticus was even kindly received by his fellow Protestant, Duke Albert of Prussia. He spent in all more than two years in the district. He eagerly followed up some early studies which Copernicus had begun on the geography of his homeland. He made a map of Prussia (which has disappeared), and he wrote a little book in praise of the beauties of that land before returning to Wittenberg in the autumn of 1541.

Rheticus brought with him (or sent from Wittenberg after returning there) a number of scientific books as a present to Copernicus. Among these was an edition

of Euclid, now translated into Latin from the original
Greek, and not merely from an Arabic version; also the
first printed Greek edition of Ptolemy's *Almagest*. When
Copernicus died he left these books, with most of his
other scientific volumes, to the Cathedral Library. Later,
however, during the Thirty Years' War, the Protestant
leader, Gustavus Adolphus, carried them off to Sweden,
with many other books and papers from that part of
the world. It has thus come about that most of Copernicus's little library is today at the Swedish University
of Upsala.

However, among all these distractions, Rheticus did
not forget what had brought him on such a long journey.
He took the earliest opportunity of getting to grips with
Copernicus's wonderful manuscript (now receiving the
finishing touches); all needed explanations he sought
and obtained from its author. Within less than three
months he had written a short summary of its contents
which he sent to his former teacher, Johann Schöner
of Nuremberg, in accordance with an arrangement already made between them.

Rheticus printed this *First Account* (as it was called
in Latin) in 1540, with the permission of Copernicus.
It was indeed the earliest reliable and detailed report
on the new system to be published. It dealt only with
a part of the Copernican theory—the part dealing with
the various motions of the Earth. The book was intended to be followed by other *Accounts*, covering the
other parts of the system; but these were never required.
For, instead, Copernicus at last sent his great book to
press. And, but for Rheticus, we might never have had it.

28. The Beginning of Immortality

Some time during the visit of Rheticus, or shortly after
his return to Wittenberg, Copernicus made up his mind
to publish his book. What finally decided him to take
the plunge, we can only guess. He may have been moved

by the entreaties, or caught something of the spirit of his hopeful young disciple, whose published *Account* had broken the ice, and prepared the way for the full revelation. Or he may have felt that his end was drawing very near; soon he would be beyond earthly praise or blame. And if the great book to which he had devoted his best years and energies were not to go down into the dust with its author, he must take the decisive step now.

In the Preface to his book, Copernicus writes of the exhortations, and even reproaches, of his friends as having at last prevailed upon him to send the work to press. Among these men he specifically names Cardinal von Schönberg and Bishop Tiedemann Giese. It was to this last-named friend that Copernicus committed the precious manuscript. Giese, in his turn, sent it for safe-keeping to Rheticus. Perhaps during the last summer holiday they had all spent together, Giese had made some plans with the young Professor as to what to do should Copernicus decide on publication.

Rheticus chose as the publisher a friend of his, John Petrejus of Nuremberg, who had read the preliminary *Account*, and who was anxious to bring out the great original. It would have been well if Rheticus could have carried out his intention of seeing the book through the press. But he had been appointed to a new post at Leipzig University, and he had to hand the task over to a local Lutheran clergyman, Andrew Osiander. This move had some rather unexpected consequences, as will be related in a subsequent chapter. However, the printing went through, and the book was published in the spring of 1543.

Copernicus lived for little more than eighteen months after Rheticus had left for Wittenberg in the autumn of 1541. Almost all we know about this last stage of the astronomer's career is found in two letters from his friend Bishop Giese.

The first of these letters, dated December 8, 1542, is addressed to one of the canons of Frauenburg, George Donner. He has not come into our story so far, but he seems to have been one of Copernicus's best friends in these closing days. Giese has heard that Copernicus has recently been taken seriously ill, and he is writing to Donner for news. What chiefly worries the old Bishop is the friendlessness of the sick man. He had always loved solitude; and latterly his brethren had shrunk from the society of a man under suspicion for his strange ideas and his friendship with Protestants. Giese, away at Kulm, vividly senses the whole situation; and he writes to Donner: "I know that he always counted you among his truest friends. I pray, therefore, that, if his occasions require, you will stand by him and take care of the man whom you, with me, have ever loved, so that he may not lack brotherly help in his distress, and that we may not appear ungrateful to a friend who has richly deserved our love and gratitude."

The second of Giese's letters was addressed to Rheticus on July 26, 1543; but by that time all was over. From the latter half of 1542, Copernicus had been repeatedly stricken with hemorrhage and apoplexy, which left him in a half-paralyzed condition. By the beginning of the new year he was believed to be at death's door. However, he lingered on until May 24, 1543—the day when the first printed copy of his book was brought to him. Giese describes the last sad scene to Rheticus: "He had lost his memory and mental vigor many days before; and he saw his completed work only at his last breath upon the day that he died."

Copernicus had been attended in his last illness by a qualified physician, Fabian Emmerich, the Vicar of the Cathedral, to whom he bequeathed one of his favorite medical books, containing many notes in his own handwriting. His other books went to the Cathedral Library, and the rest of his worldly wealth to the family of his married sister, Katherina, Canon George Donner acting as executor of the will.

When Copernicus had begun to feel his years, he had applied to the Chapter for permission to take a "coadjutor"—a younger man to help him in his duties. An arrangement of this sort was often made, with the intention that the co-adjutor should succeed to the canonry on the death of the older man. It was a way by which a canon could bequeath his place to a younger member of his family; and Copernicus proposed a son of his nephew for the office. There was some delay, but eventually his request was granted; and, after his death, his canonry duly passed to his young kinsman.

The body of Copernicus was laid to rest in Frauenburg Cathedral, though the precise location of his grave is uncertain. A memorial tablet was put up on the wall in 1581. This was removed in the eighteenth century to make way for a Bishop's epitaph; but another tablet has since been erected. The monument to Copernicus at Torun, his birthplace, dates from 1853.

29. The "Revolutions"

Copernicus owes his distinguished place among the founders of modern science almost entirely to the one great book of his which was published in the last year of his life. This book marks an epoch in the history of human thought. As a monument of scientific genius, it ranks with the *Almagest* of Ptolemy, with Newton's *Principia* and with Darwin's *Origin of Species*. The book is written in Latin, as were nearly all learned works down to about 200 years ago. This had the advantage that a scholar, having learned Latin as a boy, could correspond with other scholars all over Europe, and could read their books. To do this nowadays, he would need to learn about half a dozen foreign languages.

Before we begin to discuss special features of the book, let us try to form a rough idea of what the whole work contains and how it is arranged. At the same time we can deal with certain portions that are not of vital importance, so as not to have to return to them again.

Following the Prefaces (about which we shall have more to say in the next chapter), the work is divided up into six sections, or "Books," and these are divided again into chapters.

In the first book, Copernicus gives us his general picture of the Universe, and his arguments to prove that the Sun forms the fixed center, and that the Earth revolves round him as a planet. There is also the explanation of the seasons with which we are familiar from geography.

At the close of Book I, there are two or three chapters dealing with trigonometry. They give the rules by which, when certain sides and angles of a triangle are given to us, we can "solve" the triangle—that is, calculate the remaining sides and angles. They cover not only ordinary plane triangles, whose sides are straight lines, but also triangles on the sphere, whose sides are arcs. There is also a table of sines of angles, such as we find in a modern schoolbook (though the word *sine* is not used). Later on in his book, Copernicus makes great use of these rules and this sine-table in the calculations by which he builds up his planetary system. But trigonometry is useful in all sorts of connections having nothing to do with astronomy, for example in surveying. And it had been only recently introduced into Europe from the Arabs. Hence this portion of Copernicus's book was worth publishing separately. And when Rheticus left Frauenburg (Chap. 27), he took a copy of these chapters with him and brought them out as a little textbook at Wittenberg in the following year.

In Book II, Copernicus applies his rules of trigonometry to all sorts of problems connected with the apparent motions of the heavenly bodies on the sphere of the sky. Suppose, for example, he wanted to know the elevation of the Sun above the horizon at some given instant, as seen from a place of given latitude. Then all he had to do was to consider the triangle formed by the pole of the heavens, the zenith (Chap. 21), and the center of the Sun; upon solving this triangle he would obtain the required elevation of the Sun. All this, of course, had nothing to do with the Copernican theory. Ptolemy had dealt with the same sort of problem. Hence we shall not have anything more to say about this part of the book.

At the close of Book II there is a star-catalogue. This is a list of the brightest stars, giving the position of each on the sphere of the sky. The way this is done corresponds exactly to the way in which we give the position of places on the surface of the Earth. We say that a place has a certain *longitude* east or west of the Greenwich meridian, and a certain *latitude* north or south of the equator. In just the same way, the position of each star is given by means of a pair of measurements referring to imaginary circles on the sky. The importance of such a catalogue lay in the fact that, for the early astronomers, the stars were like fixed landmarks, or the numbered divisions of a racing-track. It was by referring to this background of stars that the positions and motions of the planets were measured. Such a catalogue has to be constructed, in the first place, by actual observations of the stars carried out with suitable measuring instruments. Copernicus, however, did not make his star-catalogue in that way. He took over Ptolemy's catalogue ready-made, merely attempting to correct certain errors into which such catalogues fall with lapse of time.

In the remaining four books, Copernicus gives detailed accounts of the supposed motions of the Earth (Book III), the Moon (Book IV), and the planets (Books V and VI). Each of these accounts leads to the construction of a *theory* (Chap. 10), that is, of a geometrical diagram showing the course supposed to be followed by that particular body. The diagram is drawn to scale; and the motion of the body is adjusted to fit the observations upon which the particular scheme is based. Each theory leads to tables serving to predict the future motion and positions of the bodies.

So far we have not mentioned the title of Copernicus's book. The fact is we do not know what name the astronomer intended to give his immortal work. The manuscript bore neither a title nor the name of its author when the latter handed it over to his friend Tiedemann Giese for publication. Both of these were added while the material was being prepared for the printer. The title given to the volume by its editors was *Six Books*

Concerning the Revolutions of the Heavenly Spheres
(*De Revolutionibus Orbium Coelestium Libri VI*). But
we shall call his work simply *Revolutions*.

30. The Prefaces

We have seen how Copernicus kept his manuscript locked
up for some thirty years. During this time he kept revis-
ing it and adding fresh observations, but he could never
feel that it was quite ready to publish. Another reason
for the long delay was that he shrank from the storm of
criticism which he knew would be unloosed from all
sides against the ideas set forth in his book.

This criticism would come from two sets of people.
On the one hand, there were the old-fashioned philosophers who insisted that the Earth was the fixed center
of the Universe, because Aristotle had said so. On the
other hand, there were the Churchmen who might ac-
cuse Copernicus of contradicting the Bible because certain texts seemed to suggest that the Earth stood still.
(Of course many Churchmen were philosophers, and
would criticize Copernicus on both counts.) In this respect the Protestants were even stricter than the Catholics. But the Catholics themselves, alarmed at the revolt
of the Protestants, were tightening up their standards
of belief and discipline. A movement had started which,
a century later, was to make possible the persecution of
Galileo for teaching the very thing that Copernicus was
now about to put forward.

Hence, when at length Copernicus had been persuaded
to publish his book, he decided to take the bull by the
horns and boldly to dedicate the work to the reigning
Pope. This was Paul III, the first of a series of much
more decent Popes who from now on sat on the throne
of St. Peter. He was a scholar, a humane and intelligent
man, and under his protection Copernicus knew that
his book and himself would be safe from persecution.
Hence the Preface to the *Revolutions* is a dedication to

the Pope, over the composition of which Copernicus took a lot of pains.

His chief excuse for putting forward his theory was, he writes, that there were already really two theories of planetary motion in the field. One was derived from Aristotle, the other from Ptolemy. As two theories were allowed, there could be no great objection to a third, especially as the other two were not very satisfactory. Aristotle's theory was supposed to follow sound physical principles; but it was vague, and had never provided tables showing the actual movements of the planets. Ptolemy's theory was ultimately the basis of the tables still in use in the time of Copernicus; but it broke the accepted laws of physics in various ways.

Copernicus tells the Pope that he was grieved at this state of things, and so he turned to the old Greek and Latin writers to see if any of them had suggested something better. He found that several of the Greek thinkers had explained the observed behavior of the heavenly bodies by supposing either that the Earth turned daily on its axis, or that the Earth revolved round the Sun together with the other planets, or that both these movements took place together. In his manuscript he mentions the speculations of Philolaus (Chap. 7), Heraclides and Aristarchus (Chap. 10). Whether or not he really drew his ideas from these old sages, he thought it well to mention them, knowing that his contemporaries had an excessive reverence for the wisdom of the ancients. He says that, following up their suggestions, he found that, by supposing ourselves to be observing the heavens from a revolving Earth, "not only would the planetary appearances follow as a consequence, but the order of succession, and the dimensions of the orbits of the planets, and the heaven itself, would be so bound together that in no part could anything be rearranged without upsetting the other parts and the whole Universe."

Another reason which Copernicus put forward for claiming the Pope's protection was that his book would help to solve the old problem of calendar reform about which he had been consulted in 1514 (Chap. 22). For the book provided an accurate value of the length of the year; and it went a good way towards clearing up the intricate motion of the Moon, upon which the date of Easter depends. It was indeed upon the improved tables in the *Revolutions* that Pope Gregory based his reform of the calendar later in the century.

We do not know whether Copernicus had received any encouragement to address himself to the Pope, perhaps from Cardinal von Schönberg (Chap. 25), whose letter occupies a prominent place at the beginning of the book. Nor did any question arise of Paul's acceptance or protection of the book. For long before it could have reached Rome, its author had passed beyond the praise or blame of any earthly potentate.

Besides the Preface we have just described, the *Revolutions* possesses, curiously enough, a second Preface, not written by Copernicus at all. It owed its appearance to the unfortunate circumstance that Copernicus was unable to see his own book through the press. The responsibility fell on Rheticus; and even he was soon called away from Nuremberg, leaving a local Lutheran clergyman, Andrew Osiander, to see after the publication of the book.

Osiander was something of an astronomer, and Copernicus had written to him asking what sort of a welcome a book about the motion of the Earth would be likely to receive. Osiander, living in the swim at Nuremberg, would be better able to answer this question than would Copernicus, marooned at Frauenburg. The reply was not encouraging. But it is a fact that the motion of a planet, *as it appears to us,* can often be represented equally well by several different theories. Hence even the Greek astronomers recognized that such a theory need not reproduce the *true* motion of the planet in space. It may be merely an artificial device serving for the calculation of tables giving the future positions of the planet. Osiander suggested that Copernicus should declare his theory to be just such an artificial device. There would then be no question of any *real* motion of the Earth, and all would be well. But Copernicus would not take this easy way out.

However, as we have said, a curious chance left Osiander, of all men, in sole control when the book was being printed. He took the opportunity to put in a short Preface of his own (but unsigned) on the lines of his previous advice to Copernicus. This really had the effect of altering the whole meaning and intention of the book. The second "Preface" was quite out of keeping with the rest of the work. But if it was a fraud, it was a very innocent fraud, though Tiedemann Giese was very angry about it. For it praised Copernicus, and therefore could not have been written by him. Shrewd readers quickly saw through the imposture, and soon the whole story leaked out. It was just a well-meaning attempt to make the book appear harmless; and, strangely enough, it succeeded in its purpose for nearly a century.

Other alterations were made by other hands than Osiander's, so that the book as first printed differed considerably from the manuscript as it left the hands of its author. In fact, the printers must have worked from a copy made by someone who thought he could improve upon the original. For many generations no one suspected this. For the manuscript soon disappeared, and it could not therefore be compared with the later printed editions of the *Revolutions* which were brought out in various parts of Europe—at Basel (1566), Amsterdam (1617) and Warsaw (1854). All these were merely corrected copies of the original Nuremberg edition (1543).

However, by a lucky chance, the manuscript turned up in the middle of the nineteenth century at Prague. Since then it has been very carefully studied, the many corrections and additions showing how slowly and painfully Copernicus built up his book. The dates of the observations which he includes in the several sections often give us some idea of the times at which those sections were written. He very likely began the book during his stay at Heilsberg, and he gave it two main revisions —the first, during his early years at Frauenburg, the second, in the lull which followed the war. He may also have given it some final touches during the visit of Rheticus about 1540.

Once the manuscript had been recovered, it was possible to print the *Revolutions* in the form in which

Copernicus intended it to be read. This was done in the edition which was published at Torun in 1873 to mark the 400th anniversary of the astronomer's birth there.

31. The Motion of the Earth

We usually associate the name of Copernicus with all sorts of startling novelties and revolutionary theories. Hence it comes as rather a surprise, in reading through his book, to find how many of Aristotle's ideas he swallowed.

For example, he believed, with the ancients, that a heavenly body must always move in a circle at a steady rate about the center. But it was by means of that very law that Copernicus claimed to be able to prove that the Earth is moving. For, he argued, if we watch a planet, it does not appear to be moving at a steady rate. This strongly suggests that the Earth, the platform from which we observe, is not fixed at the center of the planet's circle.

Is the Earth, then, moving? But if it *is* moving, how are we to know? How, in fact, *do* we know that we are moving when we are riding in a vehicle? By seeing all the objects around us apparently traveling in the opposite direction. In just the same way, says Copernicus, "if we attribute a certain motion to the Earth, it will appear as a similar but contrary motion affecting all things outside the Earth, as if we were passing them by." Hence, the only way in which we can hope to discover whether the Earth is moving is to look out for some contrary motion which appears to belong to all the heavenly bodies. One obvious example of such a motion is the daily revolution of the sphere of the sky from east to west which produces the rising and setting of the Sun, Moon, planets and stars. Copernicus asks us to believe "that the heavens have no part in this motion, but that the Earth turns from west to east."

Copernicus makes one other application of this same principle, only this time it is not quite so simple. The great difficulty in discovering one's own motion from the apparent motion of surrounding objects is that these objects may have motions of their own. It is no use trying to judge the ground-speed of an airplane from the rate at which it is overtaking clouds if the clouds themselves are moving. Now the planets have motions of their own which complicate matters. Copernicus accordingly fixes upon the annual circuit which the Sun appears to perform against the background of the stars. The ancients thought that the Sun described a circle with the Earth fixed at the center. But one could equally well assume that the Earth described a circle about a fixed Sun. In that case, what *real* motions must the planets be supposed to possess in order that such a motion of the Earth may make them appear to move as they do?

The solution of this problem was not very easy; but the answer was very simple. Supposing the Earth to revolve in a circle about the Sun at the center, then each planet must also be revolving in a circle about the same center. Thus the orbits of the Earth and planets are a system of concentric circles having the Sun as their common center. When Copernicus had arrived at this beautifully simple result, his mathematical instinct, formed in the school of Pythagoras, told him that it was the truth. "In the midst of all dwells the Sun. For who could set this luminary in another or better place in this most glorious temple, than whence he can at one and the same time lighten the whole. . . . And so, as if seated upon a royal throne, the Sun rules the family of the planets as they circle round him."

A prominent feature of Ptolemy's planetary theories had been the huge *epicycle* (Chap. 10) in which each planet moved round about its mean position. This contraption was necessary in order to explain why a planet's speed and distance appeared to vary so greatly as seen by an Earthbound observer. Copernicus had now shown that this supposed motion of the planet on an epicycle was simply an illusion produced by transferring to the planet the real annual revolution of the Earth about the

Sun. He thus abolished at one stroke the most considerable complication of the motion of each planet on the old theory.

The Copernican system, then, in its broad outlines, assumes that the Earth and the planets all move round the Sun in concentric circles. Their order of arrangement is shown in Fig. 14B, with Mercury tracing out

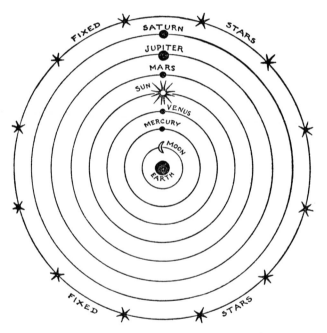

FIG. 14A. THE SYSTEM OF THE WORLD IN BROAD OUTLINES
According to the Greek and Medieval Astronomers

the innermost, and Saturn the outermost, circle. The Moon circulates round the Earth all the time that the latter is making its way round the Sun. The stars were thought to be situated at the greatest distance from the central Sun. Whether they were fixed to a sphere, or scattered through infinite space, Copernicus seems to have left as an open question.

The simple planetary theory we have just described was rather too simple to account for all the facts of observation in detail. For the planets really follow a rather more complicated law of motion than that of revolving steadily in circles. This law was not discovered until about sixty years after the death of Copernicus. Meanwhile, he tried to adjust his theory to fit the observa-

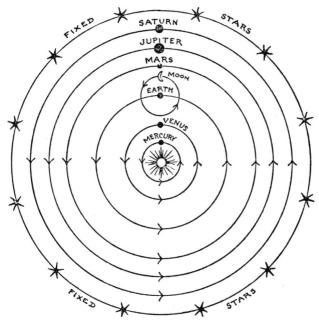

FIG. 14B. THE SYSTEM OF THE WORLD IN BROAD OUTLINES
According to Copernicus

tions by refining on the simple scheme of Fig. 14B. Into the details of these refinements we shall not enter, though a great part of the *Revolutions* is taken up with working them out. We shall merely state that Copernicus employed for this purpose the same geometrical devices as Ptolemy—the *deferent, epicycle* and *excentric* (Chap. 10).

Thus we must distinguish the Copernican *theory*—the fruitful idea of a Sun-centered Universe, which was of permanent value—from the Copernican *system*—the mass of artificial "dodges" for representing the motions of the planets which was soon to be swept away by Kepler (Chap. 38).

In order to calculate the sizes of these various planetary circles, Copernicus made use of skilfully combined sets of observations—his own and those of early astronomers. That was why he had to observe the sky for himself and why he manufactured instruments and set up his turret-observatory at Frauenburg.

Copernicus was won over to his new theory by the charm of its great simplicity. But he knew that it would be criticized by everybody as opposed to accepted ideas of mechanics, and contrary to common sense. So in his book he made a little list of the sort of objections he expected people to make, and tried to answer them beforehand.

For example, it would be argued that, if the Earth turned round once in a day, we should all be carried round towards the east at a great speed. It might be as much as 1000 miles an hour if we lived near the Equator. Anything floating in the air, such as a cloud or a bird, would be left behind, and would *appear* to us to be traveling westward at a breakneck speed. Or if a child threw a ball straight up into the air, it would come down again some hundreds of yards to westward of the place where it was thrown. Copernicus replied that the air surrounding the Earth, together with the bodies floating in it, is largely carried round the Earth in its daily motion.

Then there was the argument that, if the Earth spun round at such a terrific speed, it would break into pieces like a wheel that is made to revolve too fast. To this Copernicus replied that, if the Earth did not rotate once a day, then the sky must do so to make the stars rise and set. And, because the sky is so much bigger than the Earth, its surface must be moving proportionately faster. and therefore must be in much greater danger of breaking in pieces. So that, after all, the Earth was much more likely to be turning than the sky.

As we have said, Copernicus did not forget his old promise to Pope Leo X that he would try to determine the exact length of the year. But the task proved rather more difficult than he might have anticipated. The trouble is that there is more than one kind of year. On the one hand, there is the time in which the Sun appears to make one complete circuit of the sky against the stars and to get back to his starting point (Chap. 3). We can call that the *star year*. On the other hand, there is the period in which the *seasons* repeat themselves—say the time from the beginning of spring, 1947, to the beginning of spring, 1948. This we can call the *seasonal year*. The seasonal year is about 20 minutes shorter than the star year. This is not very much. But the difference between the results of reckoning by one kind of year or by the other adds up with lapse of time. After two years it is forty minutes; after three years, an hour; after seventy-two years, a day, and so on. Now the star year is the easier to measure; but the seasonal year is the more useful as it corresponds to the return of seedtime and harvest, and of the longest and shortest days.

However, Copernicus thought the length of the seasonal year varied very much. He was led to this wrong conclusion by studying old, unreliable observations. As we have already seen (Chap. 21), Copernicus took other people's observations at their face value, and he gave himself a lot of needless trouble in consequence. In this case he filled pages of his *Revolutions* explaining how the Earth, besides rotating once a day and revolving once a year, must also perform a sort of double wobble so as to agree somehow with some doubtful observations of the old Greek and Arab astronomers.

Still, Copernicus found the lengths of the star year and of the seasonal year pretty accurately. For example, his estimate of the star year, 365 days 6 hours 9 minutes 40 seconds, was only about thirty seconds greater than the true value. He also worked out what he took to be the complicated "law" according to which the length

of the seasonal year varied. He improved, too, on the
accepted values of the various periods connected with
the Moon. Thus he kept his promise to the Pope; and
his results were used, later in the century, in the reform
of the calendar.

32. The Size of the Universe

One of the chief problems that the astronomer is expected to solve is that of finding how far the heavenly
bodies are from the Earth, and how big they are.

The first thing that had to be done was to find out
how big the Earth is. The ancients used to measure the
elevation of the midday Sun above the horizon from two
stations, one due south of the other. From the difference
of the two elevations (taken on the same day) and the
distance between the stations, they could work out the
size of the Earth by doing a sum in proportion. This
was done in Egypt about 250 B.C.; and it gave the radius
of the Earth with a fair accuracy.

Using this radius as a yardstick, the Greeks next tried
to find the distances of the Moon and of the Sun. In
theory, their methods were sound, and they obtained a
good value for the Moon's distance—about sixty times
the radius of the Earth. But their observations were not
sufficiently refined to give them the Sun's distance. They
concluded it was about 1200 times the Earth's radius,
whereas actually the number should be about 23,000.

To arrive at the distances of the planets, the ancients
resorted to mere guesswork. For example, it was sometimes supposed that these distances stood to one another
as the lengths of a harp string giving notes which would
sound together as a musical chord. Later on it was recognized that the distance of any given planet from the
Earth varied within certain limits. There thus arose the
idea that the greatest distance of any planet from the
Earth must be equal to the least distance of the planet
next outside it. Thus the greatest distance of Mars must

be equal to the least distance of Jupiter. The idea was
that nature would never allow useless, empty spaces in
the Universe.

Copernicus looked into all these traditions concerning the distances of the heavenly bodies. He worked out
afresh the distance of the Moon from his own observations, arriving at a value of 60.30 Earth's radii as against
the modern value of 60.27. He repeated Ptolemy's calculation of the Sun's distance, but failed to improve upon
it. However, as regards finding the distances of the planets

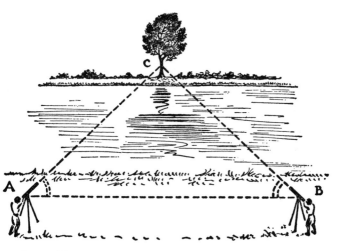

Fig. 15. Measuring the Distance of an Inaccessible Object

—the radii of the circles they describe round the Sun—
the methods used by Copernicus were a vast improvement; and his values differ but little from those accepted
today. Thus, taking the radius of the Earth's orbit as
the unit of length, Copernicus gives the radius of the
orbit of Mars as 1.520 (modern value 1.524), and that of
Jupiter as 5.219 (modern value 5.203).

The method employed by Copernicus to determine the
distances of the planets is applied in various ways in
astronomy. It depends upon a principle which keeps

cropping up in the arguments of that period for and against the motion of the Earth. Hence we had better try to explain it once for all.

We have already had one example of the principle. Copernicus used it when he argued that any motion of the Earth would make everything outside the Earth *appear* to travel in the opposite direction (Chap. 31). It is the experience we have whenever we look out of the window of a moving railway-carriage. Objects by the line-side show a steady alteration in their *direction* as seen by us. They also change their *grouping,* because the nearer an object is, the more it appears to be affected by our motion. In fact, we could roughly judge the distance of an object by finding how much its direction is altered by our traveling, say, a hundred yards.

This is actually done in surveying. A well-known problem is that of finding the distance, say, of a tree on the far bank of a river which cannot be crossed (Fig. 15). The method is to measure out a base-line AB on the near bank of the river, and then with a theodolite, or some such instrument, to measure the *bearing* of the tree C, first from A and then from B, that is, to measure the angles BAC and ABC. Knowing these angles and the length AB, we can "solve" the triangle ABC and find the distance of the tree from A or B.

Let us think of A and B as two positions of the Earth in its orbit on two different days of the year, and C as the position of a planet. Then that will give us a rough idea of the way in which Copernicus made use of the motion of the Earth to determine the distances of the planets. Of course, his problem was more complicated because the planet is moving as well as the Earth; and it is the planet's distance from the Sun and not from the Earth that is wanted. But the same principle holds good, that the nearer a thing is, the more its apparent direction is altered by a given shift in the position of the observer. Such a change in the apparent direction of a thing produced by the motion of the observer is called *parallax.*

But there was another way in which the same principle could be applied. If the Earth really traveled round the Sun once in a year, this motion ought to produce an apparent shift in the positions of the *stars*. It might well be less noticeable than for the planets, because the stars are farther away; but it would be worth looking for. Now, in the time of Copernicus, nobody had ever been able to discover such an annual shift, or parallax, in the stars; and that was one of the main reasons (and the soundest reason) why astronomers had always maintained that the Earth did not move.

Copernicus surmounted the difficulty, for the time being, by saying that the stars were so far away that the shift they showed was too small to be noticed. That seemed a reasonable excuse in those days of crude instruments. But the years passed, and, with the introduction of the telescope, methods of observation became more and more refined, yet still no parallax of the stars was observed. Evidence kept accumulating in favor of the Copernican theory; but the absence of star-parallax remained a black spot until just over a century ago, when minute shifts were detected in several of the nearest stars, and the Copernican theory was finally established.

33. Copernicus the Man

When we turn from tracing the life-story and exploring the writings of Copernicus the astronomer to attempt some assessment of him as a human being, we seek in vain for the sort of material upon which such judgments are wont to be based. We have nothing in the way of a biography of the man, written by one who knew him personally, and containing those little revealing touches which often help us to understand the great ones of the past. Even more serious is the almost complete disappearance of the correspondence of Copernicus—the letters he exchanged with his family and friends, or with the learned men of his day. This correspondence was long preserved at Frauenburg; but at some time or other it shared the common fate, or was perhaps filched by some Copernican scholar whose zeal outran his honesty.

We turn to the various portraits of Copernicus which have come down to us and for whose authenticity varying claims are made. These portraits bear a general likeness to one another. They reveal a man in the prime of life, of somewhat harsh features, with thick, short hair, a brooding expression, and a penetrating glance.

Perhaps we can learn most about the character and disposition of Copernicus from his life-story itself, studied against the background of his age. In our day a man usually makes a name for himself by excelling in some limited sphere of activity, such as scholarship, soldiering or politics, art, money-making or sport. Few people think any the less of such a man if he is grossly ignorant concerning everything outside his chosen field, and almost inarticulate even about his special interest in life. But when Copernicus lived, a different standard prevailed, especially in Italy, where he spent some of his most impressionable years. There a man was judged by the range and balance of his interests and attainments, by his ability to give distinguished artistic expression to whatever was in him, by his wisdom in the councils of the state, by his contributions to the useful arts, by his physical vigor manifested in sport and war. This ideal of manhood was largely realized in such men as Leonardo da Vinci and Philip Sidney. Its inspiration can be traced in the character and career of Copernicus as we recall some of the aspects of his many-sided achievements.

To what we have already written of his labors in astronomy it may be added that he had formidable difficulties to overcome in this science. For example, the old observations were dated according to calendars whose relations to the Christian calendar were somewhat obscure. All that had to be cleared up so that the exact number of years, days, hours, etc., between, say, two eclipses of the Moon could be known. Again, the modern astronomer can lighten the burden of computation by using logarithmic and other such tables, or by operating a calculating-machine. But Copernicus had to work out every problem using only the clumsy arithmetical methods available in his day. He was also without those invaluable aids to precision in astronomy, the telescope and the pendulum-clock.

In mathematics, he was something of a pioneer. The chapters on trigonometry occupy only a few pages of the *Revolutions*. But they contained matter which would be new to many European mathematicians; and their importance amply justified their separate publication by Rheticus.

Balancing his devotion to astronomy and mathematics, the other interests of Copernicus reveal the versatility so much admired among the men of the Renaissance. He was a linguist knowing at least Polish and German (and probably Italian) together with Latin and Greek (and perhaps Hebrew). He was, indeed, one of the few Greek scholars of his age in the part of the world where his later years were spent. His attainments in theology and Church law, and in medicine, would have carried him high in either profession had he not valued his freedom and independence more dearly than wealth or preferment. It cannot be pretended that he was filled with the heroic devotion to his sacred calling which has inspired the saints in all ages. Nowadays, he would be considered a more suitable candidate for a college fellowship than for Holy Orders. But in his own age he must have passed for an exceptionally favorable specimen of the priesthood.

Another phase of the genius of Copernicus is revealed in his contributions to economic theory, and in his activities as an administrator. His literary powers found scope in his translations of the poems of Theophylactus; they are revealed more impressively in the excellent diction of the *Revolutions*. He may well have been something of a poet; though the Latin poem of the Seven Stars, dealing with the early life of our Lord, and once attributed to Copernicus, is now thought to be by another hand. As a young man he dabbled in geography, later handing his work over to Rheticus to finish. Copernicus also won a measure of fame as an artist, and he is said to have painted his own portrait; though only a copy of this work remains, decorating the astronomical clock of Strasbourg Cathedral.

This universal culture which Copernicus absorbed during his long student years did not make him a bookworm, or a walking encyclopedia, or a mere dreamer. His mind was no scrapbook, but a chest of tools tempered by philosophy and sharpened by intercourse with the choice spirits of many nations. He returned from Italy equipped not only with the learning, but also with the knowledge of human nature and affairs which he would need to play the active part in peace and war assigned him by destiny. The whole story of the man speaks to us of his independence of mind and self-reliant tenacity of purpose. Unmoved by the prejudices or the convictions of those around him, without the solace and counsel of like-minded friends or the stimulus of active controversy with opponents, he pursued his solitary purpose for a generation, seeing it brought to fruition only in the last hours of his life.

It is natural in these days that attempts should be made to relate the achievements of Copernicus to the circumstances and needs of the society in which he lived. For example, there is an obvious significance in the stirring of interest in astronomy during a great age of ocean navigation. Other such cross-links between the astronomer's career and his social background may have suggested themselves to the reader. But the hour awaits the man; and, when all has been said, the secret behind the achievements of Copernicus remains part of the mystery which enshrouds the genius of all the great pioneers who have turned human thought into new channels.

There has been much argument as to whether the Koppernigks were Poles or Germans. Generally speaking, Polish scholars claim the great astronomer for Poland, while the Germans maintain that he was one of themselves. For Copernicus himself and his contemporaries the dispute would have had very little meaning. The division of the seamless garment of Christendom into separate nations or sovereign states, which today we accept as a matter of course, was then only beginning. And the idea that people belong to all sorts of different *races*, with boasts of racial "purity" or "superiority," is something that has come up practically within the memories of people still alive. Hence if the astronomer had been asked concerning his race and nation, he might have replied that he was a loyal son of the Church, but otherwise would scarcely have understood the questions.

Even for us today it is not an easy question to answer. Torun was founded by Germans; its leading citizens, like those of Cracow, were mostly Germans. Hence the astronomer may well have been of German extraction. The possible connection of the family on both sides with Silesia does not prove much either way for its population was a mixed one. On the other hand, the ancestors, especially on the father's side, must have lived for so many generations under allegiance to the King of Poland as to be, for all practical purposes, Poles. And Copernicus followed the family tradition in siding with the Poles against the Germans in times of crisis. In any case, it was Poland, and Cracow above all, that first nourished the youthful genius of Copernicus. And since his death it is chiefly the Poles who have gloried in their share in him, and have cherished the renown his achievements have brought to their heroic and ill-starred nation.

PART THREE

The Triumph of the Copernican Theory

34. The First Reaction

In these concluding chapters, we shall try to round off the story of Copernicus by giving some account of how his main ideas have come to be accepted as true. It is, of course, impossible to say how long this process took to accomplish.There were some people who accepted the Copernican theory right away. On the other hand, there may still be some cranks who maintain that the Earth is fixed in space and that the heavenly bodies revolve round it. But we shall not be far wrong if we assume that the last traces of serious opposition were swept away by Newton. And as his great book appeared in 1687, we may, perhaps, take one hundred and fifty years as roughly the length of this important stage in the growth of astronomy.

That the process took as long as this was due mainly to two causes. In the first place, we cannot prove the Copernican theory to be true in the simple, straightforward way in which we "prove" a proposition in Euclid. It was a matter of interpretation and of the piecing together of indirect evidence. On the other hand, the idea of a moving Earth conflicted both with the prejudices of common sense and with religious beliefs expressed in terms of the older view of the world.

Into this period of about a century and a half great events were crowded, giving it something of the quality of a breath-taking drama moving relentlessly to its appointed fulfilment. And as we follow the unfolding of the plot, we shall have to record, not only the gradual triumph of the ideas of Copernicus, but also their development and transformation in many ways that he could not have foreseen, and would not always have liked if he had.

The *Revolutions*, when it was first published, caused but little excitement. Very few people could understand it. Those who could were mostly astronomers. And they saw in the book merely a new way of constructing planetary tables, which was what Osiander in his Preface had declared it to be (Chap. 30). The Catholic authorities took no action against the Copernican theory, as such, for about seventy years. But the Protestants, beginning with Luther, criticized it sharply, as we have seen. They had revolted against the authority of the Pope and the Councils of the Church to take their stand on the words of Scripture. So they were very nervous about any teaching that seemed to contradict the Bible. They brought forward texts that seemed to imply that the Earth stood still while the Sun moved across the sky—for instance, that the Sun "rejoiceth as a strong man to run a race" (*Ps.* XIX, 5); that it "returned ten degrees" on the sun dial of Ahaz (*Is.* XXXVIII, 8); and that Joshua commanded the *Sun* (not the Earth) to "stand still" (*Jos.* X, 12-13). But in 1543 Luther had but three years to live. And the Protestants, being themselves a new movement which had come into being through a revolt against authority, were not in a very strong position to check other new movements.

In fact, curiously enough, some of the earliest disciples of Copernicus were Protestants. The first disciple of all, Rheticus, was, as we have seen, a professor at the Protestant University of Wittenberg. Melanchthon had founded two Chairs, or professorships, of Mathematics at Wittenberg. The junior professor was Rheticus. The senior Chair was occupied by one Erasmus Reinhold; and he, too, in a different way, played a part in establishing the Copernican theory.

Already, in a book brought out in 1542, Reinhold refers to Copernicus (then still alive) as a "second Ptolemy." Whether, in fact, Reinhold was really a Copernican, we cannot be sure. It was part of his duties to lecture on astronomy at the University; and of course he was expected to teach the Ptolemaic system of astronomy. The textbooks that he edited for his students and that have come down to us were also treated from the same point of view. Hence, if he accepted the Copernican theory, he must have kept his opinions to himself. But at least he understood the advantages of the Copernican scheme as a basis for constructing planetary tables. And in 1551 he brought out a set of tables based upon those in the *Revolutions*. The cost of printing them was paid by Duke Albert of Prussia (the former Grand Master), in whose honor they were called the *Prussian Tables*. Reinhold enlarged the tables of Copernicus and corrected a lot of slips. But even the revised figures suffered from the same drawback as all other such tables down to that period—they were based upon too few and too crude observations.

However, with all their faults, the *Prussian Tables* were more accurate than any that had gone before. And they remained unsurpassed for nearly 80 years; in fact, until the appearance of Kepler's set, based upon a completely reformed theory of the planetary motions. The *Prussian Tables* did not, indeed, owe their superiority directly to their being based upon the "Sun-centered" theory rather than the "Earth-centered." But the fact that they were superior brought them into general use, and advertised the Copernican theory in a favorable light.

35. The Copernican Theory in England

Copies of the books of Copernicus, Rheticus and Reinhold, circulating all over Europe, made the Copernican theory generally known among those interested in such matters. Nowhere was its influence more immediately and remarkably shown than in England. A greater freedom of opinion prevailed there than on the Continent. The influence of the earlier Greek thinkers had not been so completely overshadowed by that of Aristotle in England as elsewhere in western Europe. And just at this period there were several brilliant writers on scientific subjects, some writing in Latin for scholars, but some also in English for the common people. It is in the works of these men that we have to look for the earliest references to the Copernican theory in England.

In the first place there was Robert Recorde, the royal physician to Queen Mary, that daughter of Henry VIII under whom England returned in some measure to the Catholic faith. Besides being a doctor, Recorde was one of the few Greek scholars of his day. He was also a brilliant mathematician; and his knowledge of the Greek tongue enabled him to read some of the great old mathematical and astronomical books in the original Greek. Add to all this that he was one of the pioneers in modern methods of teaching science to beginners. Recorde turned all these qualifications to good account by writing a set of textbooks on mathematics. They were meant to be of use to intelligent artisans who knew no Latin, and they were therefore written in the picturesque English of the time. They also formed a complete series, beginning with easy arithmetic, and going on through Euclid and algebra to astronomy. The books were made more readable by being written in the form of dialogues between a Master and a Scholar; and they bore quaint titles.

The dialogue dealing with astronomy was called *The Castle of Knowledge*. Here the Master gives Ptolemy's arguments that the Earth is at rest at the middle of the Universe. He admits that other men have been of a different opinion, among whom he names Philolaus (Chap. 7) , Aristarchus (Chap. 10) , and Copernicus, "a man of great learning, of much experience and of wonderful diligence in observation," who has declared "that the Earth not only moveth circularly about his own center, but also may be, yea and is, continually out of the precise center of the world 38 hundred thousand miles." The Scholar is inclined to ridicule this view; but the Master tells him the matter is beyond his present understanding, and bids him "condemn nothing that you do not well understand." He promises, on some future occasion so to set forth the view of Copernicus "that you shall not only wonder to hear it, but also peradventure be as earnest then to credit it as you are now to condemn it." So, reading between the lines, perhaps we can hail in old Robert Recorde the first Copernican of the English-speaking world. The date of his book is 1556, or thirteen years after the *Revolutions*.

There were other writers of the period who were led to take an interest in the Copernican theory through the use they made of the *Prussian Tables* of Reinhold (Chap. 34). Among these was the Yorkshireman, John Field, who, in the same year 1556, published an Almanac giving the places of the Sun, Moon and planets for the ensuing year. He praises the "Herculean labors" of Copernicus; and he justly claims that his own almanac, based on Reinhold, is a great improvement upon the others then in use in England. But he does not say whether he accepts the Copernican theory as the really true picture of the Universe.

Some people thought it quite likely that the Earth turned round daily on its axis, but were not so sure about its revolving yearly round the Sun. We find this attitude in another English royal physician of the sixteenth century—William Gilbert, Queen Elizabeth's doctor. Gilbert was a pioneer in a movement which became steadily more important during the period we have still to cover in these pages, and which has since gone from strength to strength. This movement was nothing less than the rise of experimental science, the practice of seeing for ourselves how things behave under conditions that we can control, instead of arguing about how they ought to behave. Gilbert applied this method of experiment to the study of magnetism and electricity about which little was known in his day. His great book, setting out the results of nearly twenty years' work, was published in 1600.

The magnetic compass needle was indeed known in those days and was used by sailors. Gilbert explained its properties by supposing that the whole Earth was one huge magnet, to the two poles of which the needle always pointed. He performed the experiment of floating a magnet on a little raft on water, and watched it spin round under the attraction of the Earth so as to point north and south. He argued (mistakenly) that the whole Earth, being one great magnet, must also have a natural tendency to spin round on its axis, and that accounted, he thought, for the daily rotation of the Earth according to Copernicus. He also thought there was a good reason why the Earth should keep continually turning. If it did not do so, the side of the Earth which was always turned towards the Sun would be scorched, and the side turned away would be frozen. Or else, as Ptolemy supposed, the whole heaven would have to turn round once every day while the Earth stood still. And Gilbert thought this unlikely; for how could the tiny Earth resist the mighty swirl which carried all the stars and planets round once in a day?

The Earth's yearly revolution round the Sun seemed to have no obvious connection with magnetism. Accordingly, Gilbert had little faith or interest in that part of the Copernican theory. But, by starting the idea that it is magnetism that keeps the heavenly bodies moving, he exerted an important influence on Kepler (Chap. 38) and other thinkers of the seventeenth century.

36. *The Unbounded Universe*

Men have often wondered how big the Universe is. They have asked, Does space extend without limit in all directions? Or is it bounded by some kind of wall; and, if so, what is on the other side of the wall? Are there other Universes such as ours? The earliest Greek thinkers asked such questions. And Plato and Aristotle replied that our Universe is a sphere of limited size, and that there is nothing whatever outside it, not even empty space.

For Aristotle, indeed, the limited size of the Universe seemed to follow necessarily from his belief that the heavens turned round completely once every day. For if any part of the heavens were infinitely distant from the

center about which the whole turned, then that part
would have an infinite road to travel, and would have to
move at an infinite speed in order to cover it in twenty-
four hours. And Aristotle declared that there could be
no such thing as an infinite speed; so the Universe could
not be infinite. This view generally prevailed throughout
the Middle Ages; and Copernicus himself did not try to
upset it. In his diagram of the Universe, upon which our
Fig. 14B, is based, there is an outermost circle labeled
"the motionless sphere of the fixed stars." Whether that
was the limit of the Universe or not, Copernicus was con-
tent to leave as an open question to be debated by "philosophers."

However, one result of the revival of ancient learning
at the end of the Middle Ages was that people rediscov-
ered the old Greek questionings about whether the Universe was infinite. And Copernicus, by transferring the
daily rotation from the heavens to the Earth, destroyed
the force of Aristotle's argument. For there was no reason
why the Universe should not extend without limit in all
directions provided it was not expected to turn round
once in a day. In fact, one of the most striking developments of the generations immediately following Copernicus was the growing tendency to regard the Universe
as infinite.

This old idea seems to have been revived by another of
the early English Copernicans, a distinguished scientific
writer called Thomas Digges. He published an almanac
in 1576 containing, among other things, an explanation
and defense of the Copernican theory, and an English
translation of an important chapter of the *Revolutions*.
There is also a diagram of the Copernican universe—the
first of its kind to appear in an English book. And, what
is of even greater interest, this diagram shows the fixed
stars as no longer confined to a sphere. Instead, they occupy a region extending outwards in all directions to an
infinite distance from our system. That system, according
to Digges, still retained its position at the center of the
scheme of things. But an infinite space cannot truly be
said to have a center at all. And so the next stage in the

process was to dethrone the Sun and his train of planets from their position of privilege, and to make them just one of an infinite number of similar systems floating about in a boundless space. The man who took this bold step was Giordano Bruno, the runaway monk whose tragic fate marked the close of the century of Copernicus.

Born near Naples in 1548, Bruno joined the Dominican Order while still a boy. But he revolted against the teachings of the Church on a variety of subjects, and he was forced to fly from Italy. Thereafter he lived the life of a wandering teacher and writer until, lured back to his native land, he was brought to trial for his many departures from the faith. After a long imprisonment, he was burned at the stake in Rome in 1600.

Bruno was an ardent Copernican. But he revolted against the world-picture of Aristotle in a much more thoroughgoing fashion than the astronomer-priest had ventured to do. For Bruno, space was infinite; the Sun was merely a star; the stars were suns, each with its own train of planets, all moving freely like living things. The pattern of our solar system was thus endlessly repeated through space like the design on a wallpaper. Bruno's chief argument for a boundless universe was that God's powers are unlimited, and therefore they must find expression in an infinite work of creation. Otherwise His capacities would be only partially realized. He began to publish dialogues setting forth these ideas in 1584, during the two years that he lived in London; it is possible that he met and was influenced by Thomas Digges at this period of his life. He guessed at many things that science has since brought to light, that the stars are not fixed in a sphere; that everything is made of the same sorts of matter; that the solar system contains planets undiscovered in Bruno's day; that a comet is a sort of planet, and so forth.

Giordano Bruno did a great deal to make the Copernican theory generally known on the continent of Europe. His idea that the Sun and stars were inhabited by beings like ourselves was widely accepted until about 150 years ago; it has declined following the discovery that these bodies are made of white-hot gas. It would not be true to say, as is sometimes said, that Bruno was burned alive for being a Copernican. Nevertheless, his fate chilled the hearts of other supporters of the theory, and checked its spread for a time in Catholic countries.

37. A New Star and a Great Observer

It sometimes happens that a faint, almost unnoticeable star suddenly flares up to rival in brilliance the night-suns Sirius and Vega, or the planets Venus and Jupiter. Such an event has occurred several times during the present century, giving rise to what is popularly called a "new star." Whenever anything of this sort happened, the early astronomers used to say that it could not be a star, because Aristotle had declared that a heavenly body could not possibly suffer any change. It must be some sort of explosion in the atmosphere, perhaps some inflammable vapor catching fire not far from the Earth's surface.

In the autumn of 1572, one of these new or (as we prefer to call them) temporary stars flashed out in the northern sky. The event was of historic importance. In the first place, the star was the brightest object of its kind of which we have any record, being visible in broad daylight. Secondly, the interest it aroused started developments which greatly helped towards establishing the Copernican theory.

All over Europe astronomers hailed the appearance of this brilliant object. Then they set to work to measure how far away it was. For this purpose they made use of the principle of *parallax* that we have already explained in Chapter 32. They argued that, if the object was near the Earth, then as it passed across the night sky it ought to shift its position in relation to the stars which appeared to surround it. Even the Moon shifts through a whole degree as it passes from the zenith to the horizon. If the new star were nearer us than the Moon (as was thought likely), then it should show a greater shift than one degree, and this would be easily measurable. But the most skilful astronomical observers of the period (among them the Englishman, Thomas Digges) could detect no shift at all. This meant that the new star was much farther away from us than the Moon. And since it did not share the motions of the planets, it must belong to the region of the fixed stars where (it was believed) no changes could take place, and nothing new could come into being. This contradiction of one of Aristotle's fundamental principles helped towards the overthrow of the older astronomy, and the rise of the modern science of the heavens.

However, the new star of 1572 has a second claim to a place in our story. Its dramatic appearance roused to activity the man who proved, next to Copernicus, the greatest astronomer of the sixteenth century. This was Tycho Brahe, the Dane. In some respects, notably as an observer, Tycho was perhaps an even greater man than Copernicus. He certainly enjoyed more favorable conditions for his work. However that may be, Tycho made an indispensable contribution to the triumph of the Copernican theory even though, strangely enough, he did not believe in it!

Born in 1546, the son of a nobleman, Tycho was sent the round of the northern universities to prepare for a career in politics. But he neglected the subjects he was supposed to be studying in order to devote his days and nights to astronomy. He had only crude, home-made instruments; but they sufficed to show him how inaccurate were the planetary tables of the period—even the *Prussian Tables* (Chap. 34) based on the work of Copernicus. After his return to Denmark, Tycho drifted into the study of chemistry, and he fitted up a laboratory in an outbuilding of the house where he was staying. It was while returning thence one November evening in 1572 that he cast his eyes up to the sky and beheld the brilliant new star.

Tycho played a leading part in the attempts to estimate its distance, showing how the results generally went to prove that the object belonged to the heavens and not to our atmosphere. In after years he showed that comets, which had likewise been regarded as mere atmospheric explosions, were in reality heavenly bodies. And he pointed out that, since comets wander about freely in regions supposed to be occupied by solid spheres carrying the planets round, therefore these spheres could not have any real existence.

The fame of Tycho Brahe came to the ears of the King of Denmark who established him in a magnificent observatory on an island in the Danish Sound. Here he labored for over twenty years, helped by his growing family and by a band of assistants. His special task was to observe the Sun, Moon, and planets in their courses round the sky, recording their positions at frequent intervals over many years. He knew that only by comparing together a great number of such recorded observations would it be possible to construct an accurate planetary theory (Chap. 10), and to calculate reliable planetary tables. The great weakness of the Copernican system, and of the tables calculated from it, was that it was based upon so few observations. Even of these, some were probably spurious, and none had been made with accurate instruments. Tycho planned and constructed his own instruments; his equipment and methods of observing were a great advance on those of earlier astronomers.

Tycho Brahe had a deep respect for Copernicus and admired his system as a beautiful piece of geometry. But he could not bring himself to believe that "the gross, slothful body of the Earth" could possess any motion of its own. He was driven to this view, both by the principles of physics accepted in those days, and by the authority of the Bible, which seemed to teach that the Earth was at rest.

Then there was the old question of parallax—the annual apparent shift in the star-places which should be produced by the Earth's revolution about the Sun (Chap. 32). Tycho searched in vain for this effect. The absence of it meant either that the Earth stood still, or that the stars were at an almost unbelievably great distance. And Tycho, influenced by the ideas of his day, could not believe that nature would allow such a vast, useless space to exist between the most distant planet of our system and the stars. Again, in those days it was thought that a star showed as an appreciable disc representing a measurable angle at the observer's eye (Fig. 11). For a star to be sufficiently far away to show no parallax and yet to be big enough to appear as a disc, it would need to be very

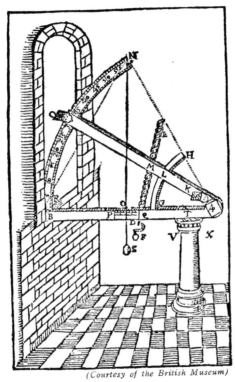

(Courtesy of the British Museum)

FIG. 16. TYCHO BRAHE'S INSTRUMENT FOR
MEASURING CELESTIAL ALTITUDES

The two wooden arms are jointed together. The lower arm is fixed horizontally; the upper arm is raised by turning the screw until the two sights are directed towards a star whose altitude above the horizon is then read off the graduated scale.

big indeed—perhaps bigger than the whole orbit of the Earth round the Sun. This was held to be an incredible size for a star; and Tycho thought it more reasonable to

reject the Copernican theory altogether, and to propose
one of his own. He supposed the planets to revolve round
the Sun, while the Sun and Moon revolved round the
Earth, each in its own proper period. In addition, the
whole heavens turned round the Earth from east to west
once every day. Some people who were dissatisfied with
the Ptolemaic theory, but who were not quite prepared
to take the plunge and embrace the Copernican theory,
adopted this so-called "Tychonic" system as a halfway
house.

The observations of Tycho and his staff, continuously
carried on for many years, were far more numerous and
accurate than those of any earlier astronomers. And it was
his intention to make use of them for working out his
system in detail as Ptolemy and Copernicus had worked
out theirs. But he quarreled with the Danish government,
and left Denmark to spend his last days in Prague, under
the patronage of the German Emperor. And before he
could get his work going again, he was carried off, in
1601, by a sudden illness. However, during these last
months at Prague, Tycho Brahe secured as an assistant
a young German astronomer, John Kepler; and it is his
name that is associated with the next decisive step in the
establishment of the Copernican theory.

38. How a Planet Moves

John Kepler was born in Germany in 1571. His father
was a drunken swashbuckler, who, though supposed to
be a Protestant, hired himself out to fight for Spain and
Rome under the notorious Duke of Alva. His mother was
an innkeeper's daughter who was later brought to trial
as a witch. When the father came out of the army, he settled down to keep a public house. The boy's schooling
was interrupted both by his having to help at home, and
by the ill health which was to dog him through life. However, he won a scholarship to the great Protestant University of Tübingen, and there began to train as a clergyman.
At Tübingen, also, Kepler came under the teaching of

Michael Maestlin, the Professor of Mathematics there. Maestlin had played a notable part in proving that the "new star" of 1572 really was a star (Chap. 37). He was a believer in the Copernican theory; and he won over the young Kepler to the same opinion. After graduating, Kepler was appointed a teacher at Graz, in Styria. Here he enjoyed considerable leisure; and he devoted it largely to astronomy.

Like Copernicus, Kepler had been influenced by the revival of Pythagorean ideas in southern Europe (Chap. 7). He thought God had created the world according to a simple pattern which we might hope to discover with a little patience and ingenuity. This pattern might (he thought) show itself through simple rules connecting the numbers and periods of revolution of the several planets and the sizes of their orbits. Kepler's first attempts to discover the secret of the Universe along these lines produced a fanciful scheme of things of which he was extremely proud, although, as it turned out, he was following a false trail. However, his first book of 1596, in which the whole thing was set out, served to make him known to the Danish astronomer, Tycho Brahe, who, as we have seen, was just about to leave Denmark to settle in Prague. In 1599, Tycho had arrived there, and he was looking out for a new astronomical assistant. Also, by that time, Kepler was looking out for a new assignment. For at Graz the authorities were now enforcing the Catholic religion; and Kepler was a Protestant. To escape persecution and worse, he was glad to take service under the Danish astronomer. That was how Tycho Brahe's unrivaled series of observations of the planets came into the hands of the one man in the world, perhaps, who could have unraveled a new planetary theory from them.

Kepler tried to reduce the observed motion of the planet Mars to a rule. He began while Tycho was alive, and carried on as his successor after his death. He tried at first to represent the planet's motion by means of the same geometrical "dodges" earlier astronomers had employed—the excentric circle and the rest (Chap. 10). By these means he succeeded in fitting the theoretical to the actual motion of the planet so closely that they never disagreed by more than about a quarter of the apparent breadth of the full Moon. But Kepler thought so highly of Tycho's accuracy that he could not imagine him being wrong by even so much as that. So he scrapped all the work he had done on these lines, though it had meant four years of tedious calculation.

In his next attempt to solve the problem, Kepler brought in all sorts of physical ideas as to how the Sun might drive the planets round in their orbits. Like Aristotle and the other old writers on mechanics, Kepler thought that a moving body would stop unless it was being continually pushed along. He thought that rays of force came out from the Sun like spokes from the hub of a wheel, and that, as the Sun turned round on itself, these rays somehow pushed the planets along. The farther a planet was from the Sun, the weaker was the force, and the slower the planet would move. But when, assuming this law to hold, Kepler tried to fit an orbit to the observed motion of Mars, he found that this orbit could not be a circle at all, but must be some sort of oval figure. It proved to be the simplest kind of oval—an ellipse. This result Kepler later showed to hold also for the orbits of the other planets, including the Earth; and it has come to be known as Kepler's first Law of planetary motion. But he found that a planet did not move at the same speed all round its orbit; it moved in a more complicated manner, expressed by a second Law.

Kepler set forth these two Laws, and he described the roundabout way he had found them out, in another great book, the *New Astronomy*, published in 1609. It is one of the most important books in the whole history of astronomy, and one of the most difficult to read. Some years later he discovered a third Law of planetary motion. It was what he had been looking for all his life—a simple arithmetical rule connecting the size of a planet's orbit with the time taken by the planet to go once round it.

Working from Tycho Brahe's observations and his own three Laws, Kepler was able to construct a great set of planetary tables which he published in 1627, and which immediately superseded all earlier tables, including those of Reinhold (Chap. 34). By the time Kepler had finished these tables, his life was nearly over. He had been caught in the disastrous Thirty Years' War, a sort of aftermath of the Reformation; and he had to travel far afield in search of a livelihood. It was after an exhausting journey, late in 1630, that he was stricken by a fever which carried him off.

Kepler was an ardent supporter of the Copernican theory. Unless he had taken its truth for granted, he could never have discovered his Laws, nor would they have had any meaning. And in one respect he made a notable advance upon the original Copernican system of the planets. Copernicus had claimed to make the Sun the center of all things. But when he came to work out the details of his system, he referred the movements of the planets to the center of the Earth's orbit, which was not quite the same as the Sun on his reckoning. This amounted to giving the Earth a privileged position in the scheme of things, and not treating it just like any other planet. Kepler put this prejudice aside and referred the motions to the true Sun. He had a sound instinct which told him that the Sun is, in some sense, the physical cause which keeps the planets moving in their orbits.

39. The New Mechanics

One of the principal reasons why the Copernican theory met with so much opposition at first was that nearly everyone held wrong ideas as to what makes bodies move and keeps them moving or changes their motion. In fact the objection of the Catholic Church to the Copernican theory was, in the first instance, not that the motion of the Earth conflicted with the words of Scripture. It was rather that the whole thing was contrary to the principles of mechanics as laid down by Aristotle, which, as we have seen, had become part of the official philosophy of the Church (Chap. 11).

Aristotle had taught that earthly bodies fall because of a sort of desire on their part to approach as near to the center of the Universe as possible. If they are seen to move in any other way, then something must be continually pushing them along. The heavenly bodies revolve round the Earth at the center because it is their nature to do so; and so forth. Until this business of motion was cleared up, it was very difficult to understand how the Earth could be moving rapidly without anything appearing to push it along, and without our being conscious of the motion in any way.

However, by the seventeenth century, sounder views on motion had begun to prevail. And this tendency culminated in the foundation of the modern science of mechanics by the great Italian scientist Galileo Galilei. Besides clearing away mechanical objections to the Copernican theory, Galileo helped it forward in two other ways. In the first place, he pointed the newly invented telescope to the heavens, and discovered there things that had an important bearing on the controversy between the old and the new systems of the world. And, secondly, in that controversy he played a manful part as the foremost champion of the Copernican theory in his generation. It is with his contributions to that theory that we shall be here concerned, though they represent but one side of his long and adventurous career.

Galileo Galilei was born in 1564—the same year as Shakespeare—in the Italian city of Pisa, which then lay in the Grand Duchy of Tuscany. His father belonged to a noble family which had come down in the world; and the boy was sent to Pisa University to train as a doctor. He took little interest in medicine. But he distinguished himself, first, by his taste and talent for mathematics and physical experiment, and, second, by his habit of arguing with his teachers and challenging the statements that they expected him to swallow on their authority. These two qualities give us the clue to his later career. Galileo left the University without a degree in 1585; and he spent the next four years at home studying the scientific works of the ancient writers. Thereafter his life story falls into three main periods, three years as a teacher at Pisa University, 18 years at the University of Padua in the Venetian Republic, and then, from 1610 to the end of his life, as Mathematician to the Grand Duke of Tuscany, with visits to Rome and imprisonments there. His work on mechanics roughly fell into the first two of these stages in his career; his telescopic discoveries and defense of the Copernican theory fell into the third; though after his condemnation in 1632 he returned for a time to his early studies in mechanics.

Various legends have grown up about the experiments by which Galileo discovered and proved the laws of mechanics. The most important discovery that he made in that branch of science was that, once a body has been set in motion, it will tend to go on moving in a straight line at a steady speed forever. In order to slow it down, or to stop it, or to turn it aside, force has to be exerted on the body. It is true that the moving bodies we see around us soon come to a standstill if left to themselves. But that is because they are acted upon by forces of the kind we class as *friction*. Thus a bicycle can be stopped by putting on the brakes. But even without the brakes, a bicycle on a level road soon stops unless constantly propelled by means of the pedals. And it is stopped by forces of the same kind as those exerted by the brakes, only now acting on the tires and the axles, or through air resistance. Without such forces, the bicycle would go on moving indefinitely.

This view of motion has an important bearing on the problem of the movements of the planets. For it had always been thought that a planet (or the sphere that carried it round) must either move of its own accord, like an animal walking or a bird flying; or else that it must be pushed along, perhaps by an angel. But since Galileo there has been no special mystery about why a planet goes on moving once it has been set in motion. For that is what all bodies tend to do. What *was* a mystery was why the planet moves in a curve round the Sun instead of traveling off along a straight line into space. A likely explanation was that there must be some force pulling the planet in towards the Sun. And that was the line along which progress was next to be made in solving the problem of a planet's motion.

40. The Telescope on the Witness Stand

When Galileo was a student, the accepted picture of the world was, broadly speaking, that of Ptolemy, the nature of the heavenly bodies being explained according to Aristotle. The Copernican theory was known; but it was accepted by only a few people, and those mainly in Protestant countries. We do not know at what date Galileo first embraced the Copernican theory. But writing to Kepler in 1597 he speaks of himself as having "adopted the opinion of Copernicus many years ago." He adds that he has never dared to say as much in print. However, in 1609, Galileo came to hear of the telescope, which had been recently perfected in Holland, and was, perhaps, already being used for astronomical purposes in England. He thought the principle of the instrument out for himself anew, and made with his own hands a series of telescopes of gradually increasing power.

Galileo's telescope worked like a modern opera glass. Kepler soon afterwards designed an improved type which has become established. It consists essentially of two convex lenses; one of these forms an image, just as the optical lantern throws a picture on the screen; the other lens is used to examine this image just as a magnifying glass is used to examine any small object. In such a telescope, objects appear upside down; but allowance can be made for this in examining a heavenly body.

When, early in 1610, Galileo pointed his telescope to the heavens, it revealed several facts which helped to overthrow the older ideas about the Universe, and to show the Copernican theory in a more favorable light. For example, it had always been maintained that the heavenly bodies were perfect spheres, free from blemishes or irregularities of any kind. But Galileo had only to look at the Moon through his glass to see that its surface was more rugged than any landscape the Earth can show, with lofty mountains and gaping valleys. Again, Galileo was one of several observers who, at this period, discovered the existence of sunspots—dark patches that come and go on the brilliant disc of the Sun. Here, then, were two examples of "imperfection" and evidences of change on the supposedly perfect and unchangeable heavenly bodies. Galileo's opponents, indeed, argued that the Moon *was* a perfect sphere after all, the apparent irregularities of its surface being filled up and smoothed off by some glassy material invisible to us!

Once again, Galileo directed his telescope to the brilliant planet Venus, which, unlike a star, showed up as a body of appreciable size. He noticed, however, that the planet did not generally appear as a round disc, but as a moon-shaped figure, whose shape altered much as the Moon's form does in the course of the month. Properly understood, Galileo's observations of Venus implied two things. First, they proved that Venus revolved round the Sun (as in the Copernican and Tychonic systems) and not round the Earth (as in the Ptolemaic system). Secondly, they proved that Venus did not shine by its own light like the Sun, as all the planets had been thought to do, but was a dark body like the Earth (or the Moon), receiving its light from the Sun. Now if Venus was a body like the Earth, that meant that the Earth was a body like Venus. And if Venus revolved round the Sun, why should not the Earth do so too?

Perhaps the most interesting of all the discoveries which Galileo made with his telescope was that of the Moons of Jupiter. These were four little bodies revolving about the planet Jupiter much as the Moon revolves about the Earth. (Their number has been increased by subsequent discoveries from four to eleven.) Apart altogether from the interest of the thing in itself, the revelation of this miniature solar system helped forward the Copernican theory. For it showed that the Earth was not the *only* center of motion. Bodies revolved about Jupiter, and if about Jupiter, why not about the Sun? Moreover, it showed that a *moving* body such as Jupiter could have other bodies revolving about it. And this met the objection of those people who maintained that the Moon could not revolve about the *moving* Earth without dropping behind it in its motion. Thus the testimony of the telescope strongly supported the truth of the Copernican theory. And thenceforward Galileo came out more and more strongly in support of it.

41. The Challenge

Galileo's brilliant discoveries led to his being recalled from Padua to his native Tuscany. But this proved a bad move as it put him at the mercy of the Church, whereas in the Venetian Republic he would have enjoyed protection. And within a few years he was in serious trouble. His all-round attacks upon the scientific ideas of the time had made him many enemies among the Churchmen who taught these ideas in the universities. These men turned the tables on Galileo by having him charged with heresy, that is, with holding and teaching opinions contrary to the doctrines of the Church.

From about the time when Copernicus died, it had become increasingly dangerous to put forward ideas with which the Church disagreed. At first this applied only to religious ideas; the victims were men such as Giordano Bruno. But in the lifetime of Galileo an increasing intolerance was being shown to disturbing scientific ideas as well; and he was the first important victim of this new policy.

By that time it was beginning to dawn on thoughtful people that the Copernican theory was going to have far-reaching effects upon man's whole outlook on life. It was not so much that man was degraded through being dethroned from his place of honor at the center of the Universe. For, after all, in the Middle Ages, the central position in the Universe was occupied by the Devil and his angels, dwelling in the heart of the Earth. Moving the Earth away from the center towards the stars was in some sense bringing it nearer to God, and thus exalting man's place in the scheme of things. But a sphere has only one center; and placing the Earth there had seemed to emphasize the *uniqueness* of man. When the Earth became merely one of the planets, this sense of uniqueness was lost, and people soon began to ask why there should not be inhabitants, human or otherwise, living on the other planets.

A still graver step was taken when it began to be maintained that the Universe was not just a sphere of limited size, but extended to an infinite distance in all directions. Copernicus had never taught that. But, as we saw (Chap. 36), he had destroyed the great argument against the infinity of the Universe which was that the heavens turned round completely once in a day. In the universe of Thomas Digges, the solar system still occupied a central position. But Bruno had argued, quite rightly, that in an infinite universe there could be no central position at all, not even for the Sun. There could be no all-embracing sphere where God could be thought to dwell and thence to govern the world. And there could be nothing to distinguish this Earth, once the scene of the Incarnation of the Son of God, from myriads of other globes, perhaps with sinful inhabitants of their own.

Nowadays we have recovered from the shock of the teachings of Copernicus and of his yet more revolutionary disciples. We have become quite accustomed to living in a universe to whose extent no limits can be assigned. And we no longer base our convictions as to the worth and destiny of man upon the circumstances of the material world which is his home during his few mortal years. But in the seventeenth century the Church felt these implications a threat to its essential doctrines, and therefore resolved to challenge the Copernican theory in the person of its most doughty champion.

About ten years before the death of Copernicus, a group of Paris students, headed by a Spaniard, had bound themselves by a vow to go to Jerusalem and convert the Moslems. An outbreak of war prevented their departure; so they offered their services to the Pope, Paul III. He formed them into a religious Order, the

Society of Jesus, later to be known as the Jesuits. The order was organized on military lines of strict obedience to superiors and willingness to go anywhere. It was a powerful instrument for the defense and spreading of the Catholic faith. Its schools long remained the best in Europe; and Galileo had been educated in one of them. Another Church activity dating from the days of Pope Paul III was the Roman Inquisition. It was a sort of spiritual police working to detect and suppress any teachings which differed from those the Church held to be true. Both these organizations were part of the re-action of the Catholic Church to the threat from the Protestant Reformation. And both played their part in shaping the destiny of Galileo and the fortunes of the Copernican theory.

42. The Champion

All his life Galileo was a devout Catholic. It distressed him to find that the opinions to which he was irresistibly led as a scientist were condemned by the Church of which he considered himself a loyal son. Accordingly, he was forced to think out for himself the relations between science and scripture. His position was one in which Christians have been placed from time to time. For example, in the middle of the nineteenth century, difficulties were felt in reconciling Darwin's theory of evolution with the Biblical account of the creation of living things. It seemed then as if man were to be de-prived of his unique central position in the world of life, just as in the sixteenth century man's home, the Earth, had been deprived of its unique, central position in the Universe.

One way of getting over the difficulty is to say that the Bible is not a textbook of astronomy (or of biology), intended to teach us things that we can discover for ourselves. It reveals spiritual truths which we could not have found out for ourselves. But it expresses these truths in ways natural to the people to whom, and through whom, they were originally revealed. That was roughly Galileo's position. It did not upset him to find that the Bible pictured the world in ways natural to the early Hebrews.

Soon after Galileo had explained his views on this matter, his enemies seized upon a little book he had written on sunspots. They extracted from it a statement of the principles of the Copernican theory, and they submitted them to a committee of the Inquisition for an opinion. This committee reported that, to make the Sun the fixed center of the world was absurd and heretical; while to say that the Earth was not the center and was in motion was absurd and, if not heresy, at least a wrong belief. This report did not, in itself, make the Copernican theory a heresy, for it was never proclaimed so by the Pope. But it provided grounds on which Galileo, in 1616, was warned not to defend the theory. At the same time Copernicus's *Revolutions* was placed on the *Index*—the list of books that Catholics were not allowed to read without special permission—until it was "corrected." The corrections would have had the effect of reducing the theory to a mere device for calculating tables, as Osiander long since had claimed it to be.

However, some years later, when the storm had died down, Galileo wrote a great book which was really meant to be a defense of the Copernican theory. In order to keep to the letter of the command laid upon him in 1616, Galileo threw his book into the form of a debate between the supporters of the Ptolemaic and of the Copernican theories respectively.

The book is entitled *Dialogue concerning the Two Chief Systems of the World;* and it relates how three friends meet on four successive days to discuss the two rival theories of the Universe, ignoring that of Tycho Brahe. They range over all the arguments on both sides. One of the characters acts as the spokesman of Galileo himself. He puts forward the fresh evidence based upon the telescopic discoveries. The book is written with great literary skill; and it is in Italian, so that other people besides scholars could read it.

On the first day the three friends debate whether heavy bodies fall in order to reach the fixed center of the *Universe* or (as Galileo supposed) in order to reach the center of the moving *Earth*. On the second day they discuss the Earth's daily turning on its axis. The point is made that the Earth could hardly remain at rest if everything outside were being carried round in a vast daily whirl, whereas, on the Copernican view, the Earth is so small in comparison with the rest of the Universe that it could easily rotate without affecting anything outside itself. On the third day they argue for and against the view that the Earth revolves round the Sun, just as the little moons revolve round Jupiter; and for the last day Galileo provides an explanation of the tides which later proved to be mistaken.

Although the weight of evidence goes in favor of Copernicus, Galileo did not dare to press the debate to a decision. He thus managed to get the book past the Censor. But immediately it was published in 1632, Galileo was attacked by the Jesuits on the ground that he had been commanded, in 1616, not to *teach* the Copernican theory, and that he had concealed this fact from the Censor. Thus, it was argued, he had obtained permission to publish his book on false pretenses. The Pope set up a Commission which considered the book in secret and produced a record of the command not to teach. Galileo could remember only being told not to *hold* or *defend* the theory; and he had done neither of these in his book. Whether his memory was at fault, or whether the record was a mistake, or a forgery, may never be cleared up. On the strength of the Commission's report, Galileo, now an invalid of close on seventy, was summoned to Rome to face the Inquisition in the winter of 1633. He was kept in close confinement until the summer, being questioned from time to time, but apparently not physically ill treated as things went in those days. However, he was allowed none of the facilities for legal self-defense normally granted to any criminal by the courts of a civilized state. He was tried in his absence, and brought up for sentence on June 22, 1633. But by that time his age and ill health, the depression born of confinement, the prolonged mental bullying to which he had been subjected, and the confusion of ideas between the spiritual authority of the Church and the testimony of experience—all these had done their work and had reduced Galileo to the condition in which he was prepared, on his knees, to renounce the Copernican theory. The sentence of the Inquisition upon him was one of perpetual imprisonment; but this was relaxed in time to a kind of "house arrest" in which he was able to continue his scientific work, in some measure, until his death.

Galileo is remembered today, however, not because he went down at the last before the forces of darkness, but because he fought so long and so successfully against them. There is even a legend that, as he rose to his feet after his forced submission to the Inquisition, and his denial of the Earth's motion, Galileo muttered something to the effect that the Earth *does* move all the same! Whatever the truth of the story, the words have become a sort of proverb implying that "the truth is great and shall prevail," despite all foolish attempts to suppress it in the interests of this or that school of thought. So it was with the truth of the Copernican theory. By his reformation of mechanics, by his telescopic discoveries, and by his battle against the ignorance of the learned men of his day, Galileo had brought the Copernican theory nearer to its final triumph than he could have dreamed in the moment of his downfall. For the year 1642, which saw the death of Galileo in captivity, saw also the birth of Isaac Newton.

43. The Whirlpool Universe

For ages men wondered what it was that kept the heavenly bodies moving in the sky. At first they pictured them as living beings, flying like birds, or sailing in boats on a celestial river. But when the early Greeks started giving matter-of-fact explanations of nature, one of the suggestions they made was that the whole world was something like a vast whirlpool. We can understand what was in their minds by watching water running out of a bath; or a "dust devil" as it rushes along the street catching up and whirling round pieces of straw or waste paper. There are famous whirlpools in the Mediterranean; and it was natural to think of the stars and planets as carried round and round just like ships caught in these. The Earth was like the mass of driftwood and weed which always tends to collect in the center of such a maelstrom.

The later Greeks and the men of the Middle Ages abandoned such matter-of-fact explanations in favor of something more mystical. They thought the stars and planets were fixed to revolving spheres. And they believed that these spheres were kept continually turning by a desire after the perfection of God, in somewhat the same way as love rouses the lover to activity in pursuit of the beloved object.

When the age of modern science began, some 300 years ago, there was a movement back to the old view that the planetary system was a sort of machine. This was supposed to work according to laws of mechanics which, though still only imperfectly grasped, were regarded as holding good everywhere throughout the Universe. And one form which this movement took was a revival of the idea that the heavenly bodies were carried round in a whirlpool of matter, or a *vortex*, to give it its scientific name.

This step was taken by a Frenchman, René Descartes, who was among the great leaders of thought in the first half of the seventeenth century, and one of the inventors of what today we call analytic geometry. Born in 1596, and educated at a Jesuit school, Descartes spent his youth in travel and soldiering. Then he settled in Holland for twenty years to lead a life of study. His fame as a scholar reached Queen Christina of Sweden; and she invited him to Stockholm to be her tutor. He went; but the wintry climate, and the Queen's habit of taking lessons at unearthly hours in the morning, were too much for Descartes's constitution, and he died a few months after his arrival (1650).

Descartes thought that the Universe was a vast collection of whirlpools of matter, or *vortices*. Each of them occupied a definite region in space, and was surrounded by others. What gives this theory its important place in our story is that Descartes early adopted the Copernican theory. He supposed that the Sun occupied the center of one such vortex, and that the Earth and the planets were carried round it in the mighty swirl. Each star was itself a sun and the center of a vortex with its own train of planets. Descartes supposed, moreover, that the Sun and the other stars exerted a pressure outwards on the matter in the space around them. He thought that this pressure, falling on our eyes, gave us the feeling of seeing light. But he also supposed that a star could become more or less coated with dark matter. The appearance of spots on the Sun was an example of this process. If a star became completely covered up, it ceased to exert pressure, and the vortex of which it was the center collapsed under the pressure of the neighboring ones. The dead star was then caught up in some other vortex, where it became a *planet*. Or else it wandered from system to system as a *comet*. Our Earth, Descartes believed, was just such a clogged-up star.

Descartes wrote a little book explaining these ideas. He was just going to have it published when he heard about the condemnation of Galileo. He was a Catholic and unwilling to bring out anything of which the Church might disapprove. So he locked up the little book in his desk, and it was not published until fourteen years after his death. But meanwhile he had found a way of expressing his views which was less likely to offend the Church authorities. He declared that a body moved only if it changed its position in relation to the surrounding matter. But, he argued, the Earth did not change its position in relation to the material of the vortex which surrounded it and carried it along. Therefore his theory could not be condemned as involving the motion of the Earth; and therefore it could be lawfully published, as, indeed, it was in 1644. A later writer humorously asked whether a maggot firmly embedded in a cheese could be said to move when the cheese was sent by ship from Amsterdam to Batavia. Descartes presumably would have ruled that it did not move.

The account of the world given by Descartes became very fashionable, especially in France, and to some extent in England and elsewhere. And the fact that Descartes made the Sun the center of his planetary system gave a great boost to the spread of the Copernican theory in these countries.

44. Universal Gravitation

When Kepler tried to explain by mechanics how a planet moves in its orbit, he thought there must be some force constantly pushing the planet along from behind (Chap. 38). Later, in the light of Galileo's discoveries, it seemed more likely that the force was one pulling the planet in towards the Sun (Chap. 39). This clue was followed up by Isaac Newton, probably the greatest man of science of all time; and it led him to an explanation of the planetary motions that has satisfied men's minds down to our own day.

This planetary theory of Newton's was based upon the Copernican arrangement of the Universe with the Sun in the center. But it was part of a bigger theory, accounting for much more than the motions of the planets. And because this wider theory was so successful, it very soon came to be accepted by everyone; and the Copernican theory was accepted along with it. Thus, with Newton, we reach the end of a critical era in the development of human thought about the Universe. A few words about his achievements in this field may fitly close the story of the rebirth of astronomy that we have tried to retell in these pages.

Newton was born, the son of a farmer, in the sleepy Lincolnshire hamlet of Woolsthorpe. Many stories have been preserved (or invented) to illustrate his early taste for reading and mechanical invention, and the distaste for farming which decided his elders to make a scholar of him. The year 1665, in which he graduated at Cambridge, came to be called the "Plague Year" after the deadly epidemic which then swept through London and far out into the country. For fear of an outbreak, the University was closed and the students sent home. Among the evacuees was Isaac Newton, a newly fledged B.A. of 22. Marooned in Lincolnshire for months on end, he turned his thoughts to the old question, now once more stirring in men's minds, What keeps the planets in their orbits?

One autumn day, in the orchard at Woolsthorpe, the fall of an apple from a tree brought vividly before Newton's mind the mystery of gravity. The story is told by some who knew him well in later years. And a log, claimed to have been cut from the very apple tree, is preserved in a glass case by the Royal Astronomical Society of London! Whatever may be the truth of this story, we have it on Newton's own authority that, about this time, he began to wonder how far the force of gravity extends above the Earth. It is felt on the highest mountain peaks. Can it possibly extend as far as the Moon? And can the force which pulls the Moon out of its straight course be the same as that which makes the apple fall from the tree? Doubtless the force of gravity gets weaker the farther we go from the Earth. How weak has it become by the time we reach the Moon?

For the answers to these questions, Newton turned to Kepler's Laws of planetary motion (Chap. 38). He was the first to understand their inner meaning; and they gave him the information he needed. He found the law according to which the force of gravity falls off with increasing distance from the Earth. And he calculated that the pull of gravity on the Moon was just equal to the force necessary to keep the Moon in its orbit. In fact gravity *was* the force; nothing else was needed.

Newton was next able to show that a force of the same kind, exerted by the Sun upon the planets, would suffice to keep these bodies, too, in their orbits. The elliptic path of a planet had been, for Kepler, merely an unexplained peculiarity of its motion. But Newton showed that it followed of necessity from the particular law of gravity concerned.

He did not think all this out in the orchard. He had returned to Cambridge to win fame through his historic achievements in more than one branch of science. But he kept working at the planetary problem by fits and starts during about twenty years. Sometimes he would be held up by a difficulty, and would turn aside to these other researches. Pressed by his friends, he would take up the old problem again, only to lay it by in disgust when other men claimed the credit for his discoveries. Yet Newton himself would do nothing to make these discoveries known until his friend Halley, finding out about them almost by accident, made him undertake to publish his work, and kept him to his promise. After about eighteen months of concentrated mental effort, Newton produced his *Mathematical Principles of Natural Philosophy*—probably the greatest scientific book ever written. It is the foundation of all our modern textbooks of mechanics. And in it the Copernican system of the planets stands revealed as a vast machine working under mechanical laws here understood and explained for the first time. In this scheme of things, too, the mysterious wanderings of the comets were shown to be subject to the same laws of motion as the planets.

Newton, however, located the fixed mechanical center of the solar system, not at the middle of the Sun, but at what we may somewhat loosely call the "center of gravity" (or, more correctly, the "center of mass") of the whole collection of bodies making up the system. However, because the Sun is so much more massive than all the planets put together, this mechanical center is, in fact, not far from the middle of the Sun. Even this center was not to be regarded as fixed for long. Newton's friend, Halley, discovered that the stars do not remain quite fixed in their places in the sky. They must therefore be moving about freely in space in relation to one another, as Bruno had surmised. Thus, with the lapse of centuries, the familiar constellations must gradually dissolve into other forms. From this it was but a step to the assertion that our system is moving through space in relation to the stars. The Sun and planets all keep together like a squadron of airplanes in flight. The truth of this assertion was first verified by William Herschel about the end of the eighteenth century. Thus the position of the Sun could no longer be regarded as either fixed or central.

With the passage of the years, Newton had come to regard the pull of the Earth on the falling apple, or on the falling Moon, as just the most familiar example of a type of force which all the heavenly bodies exert upon one another. But he went further; he thought that these forces between the heavenly bodies must arise from a universal property of matter, by which every particle in the Universe attracts every other particle. The strength of this attraction, or gravitation, depends in a perfectly definite way upon the masses of the two particles and upon the distance between them. Starting out from this principle of universal gravitation, Newton was able to give a mechanical explanation of a vast range of facts to do with the behavior of the heavenly bodies. And later generations of astronomers have carried this process much further, so that many facts of which Newton was unaware now find a place in a vast system of mechanical astronomy.

That was the way in which the Copernican theory triumphed—not through any ingenious argument, nor, in the first place, through any new discovery confirming the motion of the Earth, but because it formed an integral part of Newton's all-embracing and all-conquering theory of the physical world. Newton's success broke down all opposition, not only in England, but also in America, where the Newtonian philosophy began to be regularly taught at Yale University from the beginning of the eighteenth century. However, as time went on, observations were made and ingenious tests devised which have helped to establish, in the light of the Newtonian system of mechanics, the motion of the Earth in relation to the other bodies of the Universe. These confirmations fall into two classes according as it was the daily rotation or the annual revolution of the Earth that was in question.

One of the earliest arguments for the daily rotation was based upon the seventeenth century discovery that a pendulum of given length beat rather more slowly as it was taken nearer to the Earth's equator, which implied a falling off in the force of gravity. This could be explained by supposing the weight of bodies near the equator to be partly balanced by the tendency to fly off into space due to their being whirled round at a rate of about a thousand miles an hour by the daily turning of the Earth. For the same reason the sea, and even in some measure the Earth itself, could be expected to bulge out at the equator and to be flattened at the poles. When, during the eighteenth century, extensive surveys proved that the Earth is indeed distorted in this way, it was naturally regarded as evidence for its rotation.

In the eighteenth century, also, it was pointed out that, if we imagine ourselves traveling northward from the equator of the rotating Earth, we come to places whose eastward speed is less and less. For these places have ever smaller circles to traverse in order to complete their journey round the Earth's axis once in twenty-four hours. Hence a long-range projectile, fired northward and possessing the eastward speed of its place of projection would gain on the eastward motion of the places it came to, and would therefore appear to be suffering an eastward deflection. On this principle, the Earth's daily motion was invoked to explain why the winds blowing northward into the temperate zone are, in fact, deflected eastward to become southwest winds, and why (applying the same principle the other way round) the southward blowing winds of the tropics are deflected westward to become the "northeast trades." The same principle underlies the laws governing the directions in which the air flows round the cyclones and anti-cyclones of the weather-maps.

If our long-range projectile were fired alternately

northward and southward and its deviations were allowed to accumulate, the path it traced out would gradually swing round. The bob of a vibrating pendulum is rather like a bullet fired backwards and forwards. And in the middle of the last century, a French scientist, Léon Foucault, made a pendulum of a leaden ball and a long wire and set it swinging from the dome of the Panthéon in Paris. He showed that the plane swept out by the wire did in fact gradually swing round about the vertical, relatively to the surrounding objects, in a way that could be regarded as a direct consequence of the daily rotation of the Earth. Another of Foucault's devices was the gyroscope; it worked on a principle which underlies the gyrocompass now increasingly taking the place of the magnetic compass in navigation. It consisted essentially of a massive flywheel mounted so as to turn freely about its center of gravity. When the wheel was set in rapid rotation, its axis maintained a fixed direction in space, thus betraying the Earth's rotation.

So far as concerns the annual revolution of the Earth about the Sun, the outstanding confirmation has been the detection, since 1838, of an annual apparent shift, or *parallax,* in certain stars. The earliest authentic claim was published in that year by the German astronomer F. W. Bessel, who had observed that the star 61 *Cygni* appeared to describe a minute ellipse (only a fraction of a second of arc in breadth) about its mean position in the course of a year. This observation, as explained in Chapter 32, provided a long-sought proof of the Earth's yearly revolution round the Sun.

Aristotle had distinguished sharply between the four earthly elements, moving naturally in straight lines, and the heavenly element, or aether, going round and round forever in circles (Chap. 9). Newton's system of nature knew nothing of this distinction. His rejection of a *mechanical* difference between Earth and Heaven was matched, in the nineteenth century, by the abandonment of the corresponding *chemical* distinction. It is now known that the heavenly bodies are made of much the same kinds of matter as those with which we are familiar on the Earth. Thus the revolution in thought begun by

Copernicus has grown through the centuries into the conviction that the material Universe is a single whole, no part of which enjoys any special privilege. The same system of law everywhere prevails, binding together all events in a network of causes and effects.

45. Is the Copernican Theory True?

The Copernican theory has gone on growing and developing through the years like a living thing. The early disciples of Copernicus, while helping to establish the "Sun-centered" universe, each altered the original Copernican scheme in one way or another. Thus the solar system which they bequeathed to modern astronomy bore only a remote resemblance to the primitive system of the *Revolutions*.

Copernicus and the men of his time thought of the Universe as a closed sphere of space containing all created things. The dispute was about whether the Sun or the Earth was to be placed at the center of this sphere. Copernicus put the Sun in that privileged position, and made the planets revolve round him in combinations of excentrics and epicycles. However, within a generation of the death of Copernicus, men had begun to ask whether space might not stretch indefinitely in all directions without any center at all. The next generation saw the end of the excentrics and epicycles, and of the perfectly circular motion of the heavenly bodies. And within about a century of Copernicus, an entirely new set of laws of motion had been proposed applicable to earthly and heavenly bodies alike. The distinction between the earthly and the heavenly was fast disappearing. Our system of Sun and planets had at first been supposed to be the only one of its kind. But soon the stars were conceived to be suns, each perhaps with its train of planets, and moving freely in space, with our Sun itself such a moving star.

All this amounted to casting away the old fixed land-marks to which earlier astronomers had been wont to refer the courses of the heavenly bodies. And the old dispute over the rest or motion of the Earth or Sun could no longer be expressed in such simple terms. Accordingly, during the past century, several attempts were made to prove that the Earth is moving, and to measure its speed in relation to space itself, by means of ingenious optical experiments.

The problem is rather like that of a passenger in a railway carriage who wants to find out whether, and how fast, his train is moving. On an English railway, little signposts are set up every quarter of a mile. Thus, with the aid of a watch, it is easy to find the speed of the train by noting how long it takes to travel from one of these posts to the next one. But if the passenger could not see anything out of the carriage window, he would not be able to say how fast the train was moving. Or if what he saw when he looked out was only another moving train on the adjoining track, then he would be aware only of the *difference* of the speeds of the two trains—the motion of one train *relative* to the other.

The scientist on the Earth is in much the same position as this passenger. His attempts to measure the Earth's speed relative to space itself have all failed. It seems to be one of the laws of nature that they always must fail; that the scientist is looking for something that does not exist. It seems there are no fixed landmarks in space, only heavenly bodies floating about, like the moving train on the adjoining track. Hence we cannot say how fast, or in what direction, the Earth is moving through space, not because we are not clever enough to find out, but because the statement would have no meaning. All that we can say is that the Earth and the other heavenly bodies are moving *relatively* to one another like the two trains. If we are to say that any one of them is "fixed," it is merely a matter of choice and convenience on our part in solving some particular problem.

So we come back at the end to what we said at the beginning in Chapter 1, that Copernicus made a *choice* and not a *discovery* when he decided to think of the planets as revolving about a central Sun. If we realize that, it will help us to answer the question we have placed at the head of this Chapter, Is the Copernican theory true?

We have first to understand what is meant by the "truth" of a scientific theory. In science we employ the word "truth" in a rather special sense, differing from its use in everyday life. In judging the truth of a theory we apply two main tests. On the one hand we expect a theory to be able to take hold of a mass of facts which seem isolated and meaningless and to bind them together into an intelligible system so that we can see the connections between them. On the other hand we expect a theory to be able to suggest lines of enquiry along which further investigations can be fruitfully made. These may lead to fresh discoveries resulting, perhaps, in the overthrow of the old theory and the setting up of a new one able to account for the new discoveries as well as for the facts previously known. For no theory in science is regarded as the final truth, only as a step, or as a useful tool, towards the attainment of a further measure of truth. These are the tests we have to apply to the Copernican theory.

It was by choosing the Sun as his fixed point of reference that Copernicus was able to account quite simply for certain puzzling facts known in his day, such as the peculiar apparent motions of the planets. By adopting the same general point of view, later astronomers have been able to explain a multitude of other facts of which Copernicus in his time knew nothing. Again, the "Sun-centered" conception of our system has made possible new discoveries such as Kepler's Laws, and the principles of mechanics, which, in the hands of Galileo and Newton, have accounted for the motion, not only of the planets, but also of all the bodies we see moving about us on the Earth. It is on such grounds that, looking back over four hundred years, we assert the scientific truth of the Copernican theory and vindicate the historic choice of its creator.

Books for Further Reading

For the benefit of serious students of Copernicus it may be stated at the outset that the standard books on his life and scientific work are, in German, L. Prowe: *Nicolaus Copppernicus* (2 vols.), Berlin, 1883-84; E. Zinner: *Entstehung und Ausbreitung der Coppernicanischen Lehre*, Erlangen, 1943; and, in Polish, L. A. Berkenmajer: *Mikolaj Kopernik*, Cracow, 1900, and *Stromata Copernicana*, Cracow, 1924. A short French summary of the former of these Polish works will be found in *Bulletin International de l'Académie des Sciences de Cracovie; Classe des Sciences Math. et Nat.*, March, 1902, 200 ff. Copernicus's great Latin treatise of 1543, *De Revolutionibus Orbium Coelestium Libri VI* should be consulted in Curtze's edition (Thorn, 1873). There is a complete German translation of it by C. L. Menzzer (Thorn, 1879).

English versions of important chapters of Copernicus's book will be found in H. Shapley and H. E. Howarth: *A Source Book in Astronomy*, New York, 1929. Annotated English translations of his *Commentariolus* (referred to in the present book as the "Little Commentary") and of Rheticus's *Narratio Prima* (the "First Account") are included in Edward Rosen's *Three Copernican Treatises*, New York, 1939.

An attempt to explain to students the contents of Copernicus's book, with some account of his career and of the previous development of astronomy, will be found in A. Armitage: *Copernicus, the Founder of Modern Astronomy*, London, 1938. For the general reader, a short, well-illustrated account of the astronomer's life and achievements will be found in *Nicholas Copernicus*, translated from the Polish of J. Rudnicki, London, 1943. Hermann Kesten has painted an impressive, if somewhat flamboyant, picture of the life and times of the great astronomer in his imaginative book *Copernicus and his World*, New York and London, 1945.

Books which may help the lay reader to build up the general historical and scientific background of the developments traced in these pages include the following: H. A. L. Fisher: *A History of Europe* (in one volume), London, 1936, Books I and II; C. C. J. Webb: *A History of Philosophy* (Home University Library), 1915; Charles Singer: *A Short History of Science*, Oxford, 1941; A. Wolf, F. Dannemann and A. Armitage: *A History of Science, Technology and Philosophy in the Sixteenth and Seventeenth Centuries*, London, 1935; Ivor B. Hart: *Makers of Science*, London, 1923; and A. Berry: *A Short History of Astronomy*, London, 1898. The following volumes in the *Pelican* series will also be found suggestive and interesting reading: A. D. Ritchie: *Civilization, Science and Religion*, 1945, and B. Farrington: *Greek Science, Its Meaning for Us*, 1944. A good beginner's book on the actual science of astronomy is R. L. Waterfield: *The Revolving Heavens*, London, 1942.

Fundamental books for serious students of ancient and medieval astronomy are P. Duhem: *Le Système du Monde* (5 vols.), Paris, 1913-17; Sir T. L. Heath: *Aristarchus of Samos, the ancient Copernicus*, Oxford, 1913; and, for the whole period covered in these pages, J. L. E. Dreyer: *History of the Planetary Systems from Thales to Kepler*, Cambridge, 1906.

For the developments traced in Part III of the present work, reference may be made to Dorothy Stimson's little book, *The Gradual Acceptance of the Copernican Theory of the Universe*, New York, 1917. The lives of individual astronomers and thinkers who played notable parts in this period may be studied in W. Boulting: *Giordano Bruno*, London and New York, 1916; J. L. E. Dreyer: *Tycho Brahe*, Edinburgh, 1890; W. W. Bryant: *Kepler*, London, 1920; F. Sherwood Taylor: *Galileo and the Freedom of Thought*, London, 1938; S. Brodetsky: *Sir Isaac Newton*, London, 1929. These will present no difficulty to the general reader. More advanced students may be recommended to consult E. A. Burtt: *The Metaphysical Foundations of Modern Physical Science*, London, 1925; and, for special aspects, E. Wohlwill: *Galilei und sein Kampf für die Copernicanische Lehre*, Hamburg, 1909-26; and F. R. Johnson: *Astronomical Thought in Renaissance England*, Baltimore, 1937.

Index

Abstemius, 55

Adolphus, Gustavus, 100

Albert of Brandenburg, Grand Master and Duke of Prussia, 87 ff., 98 f., 126

Alexander the Great, 22, 34, 36

Alexandria, 36, 42

Allenstein, 79, 87ff., 90 ff.

Aristarchus of Samos, 40, 107, 127

Aristotle, 34 ff., 39, 41 f., 44 f., 59, 106 f., 110, 126, 130, 131, 132 f., 138, 139 f., 142, 157

Astrology, 20 f., 43, 59, 70, 97

Astronomy, Primitive, 16 ff.
Babylonian, 19 ff.
Egyptian, 19, 21 f.
Greek, 29 ff., 32 ff., 36 ff., 149 f.
Medieval, 40, 42, 43, 48 f., 70

Bessel, Friedrich Wilhelm, 157

Bologna, 61ff., 66

Borgia, Cesare, 65

Borgia, Lucrezia, 65, 68

Borgia, Rodrigo (Pope Alexander VI), 62, 65

Brahe, Tycho, 84 f., 133 ff., 137 f., 147

Brudzewski, Albert, 58

Bruno, Giordano, 131 f., 144 f.

Calendar, 19, 21, 59, 85
Reform of, 85 f., 107 f., 115

Central Fire, 31 ff.

Church, Christian, 43, 44 ff., 60, 92 ff., 125, 139, 144 ff.

Clement VII, Pope, 95

Coinage, reform of, 89 ff.

Columbus, Christopher, 48, 59

Copernicus, Nicholas, *passim*
parentage and birth, 52 f.

at Cracow, 56 ff.
at Bologna, 61 ff.
at Rome, 65 f.
at Padua, 66 ff.
at Ferrara, 67 f.
at Heilsberg, 71 ff.
at Frauenburg, 77 ff.
at Allenstein, 79
as classical scholar, 75 ff.
as economist, 89 ff.
as observer, 59, 64 f., 75, 80 ff., 85, 87
as physician, 68, 71, 78, 96 ff.
his instruments, 80 ff.
his observatory, 80, 85
his system of astronomy, 59, 64, 73 f., 77, 89, 93 ff., 124 f. 158 ff.
his "Little Commentary," 74, 95
his *Revolutions*, 42, 74, 78, 103 ff., 125, 147

Corvinus, Laurentius, 76 f.

Counter-Earth, 31 f.

Cracow, 56 ff., 73 f., 76, 80

Dantiscus, John, 95, 98

Darwin, Charles, 103, 146

Deferent, 40, 113

Descartes, René, 150 ff.

Digges, Thomas, 130 f., **133, 145**

Donner, George, 101 f.

Earth, center of the Universe, 12 f., 21, 33 ff., 41 ff., 106, 125, 134 ff.
motion of, 12, 31 f., 39 ff., 45, 94, 100, 105, 107 ff., 110 ff., 115, 118 f., 124, 128 f., 143 f., 147, 151, 155 ff.

Emmerich, Fabian, 102